Advances in Oil and Gas Exploration & Production

Series editor

Rudy Swennen, Department of Earth and Environmental Sciences,
K.U. Leuven, Heverlee, Belgium

The book series Advances in Oil and Gas Exploration & Production publishes scientific monographs on a broad range of topics concerning geophysical and geological research on conventional and unconventional oil and gas systems, and approaching those topics from both an exploration and a production standpoint. The series is intended to form a diverse library of reference works by describing the current state of research on selected themes, such as certain techniques used in the petroleum geoscience business or regional aspects. All books in the series are written and edited by leading experts actively engaged in the respective field.

The Advances in Oil and Gas Exploration & Production series includes both single and multi-authored books, as well as edited volumes. The Series Editor, Dr. Rudy Swennen (KU Leuven, Belgium), is currently accepting proposals and a proposal form can be obtained from our representative at Springer, Dr. Alexis Vizcaino (Alexis.Vizcaino@springer.com).

More information about this series at http://www.springer.com/series/15228

Leonardo Azevedo · Amílcar Soares

Geostatistical Methods for Reservoir Geophysics

Springer

Leonardo Azevedo
Instituto Superior Tecnico
CERENA
Lisboa
Portugal

Amílcar Soares
Instituto Superior Tecnico
CERENA
Lisboa
Portugal

ISSN 2509-372X   ISSN 2509-3738   (electronic)
Advances in Oil and Gas Exploration & Production
ISBN 978-3-319-85088-7   ISBN 978-3-319-53201-1   (eBook)
DOI 10.1007/978-3-319-53201-1

Printed on acid-free paper

This Springer imprint is published by Springer Nature
The registered company is Springer International Publishing AG
The registered company address is: Gewerbestrasse 11, 6330 Cham, Switzerland

Leonardo Azevedo *dedicates the book to his family, friends and all professors who inspired him.*
Amílcar Soares *dedicates the book to his students.*

Preface

The main motivation for writing this book is to report on an existing repertoire of geostatistical methods for handling the integration of geophysical information in reservoir modeling and for presenting the successful case studies that validate them.

Geostatistical methods were introduced in the 1960s (Matheron 1965) as tools for coping with large amount of similar data and to characterize, for example, grades dispersion in mineral deposits (David 1977; Journel and Huijbreghts 1978). When geostatistical methods become popular for oil reservoir characterization in the 1980s (Deutsch and Journel 1992), the lack of well data made it necessary for another type of data integration in geostatistical methods. Hence, a new paradigm—one based on data integration—developed in geostatistical methods though joint simulations and stochastic sequential simulations with soft data.

Since the beginning of this century, oil and gas discoveries have been mainly in deep and ultra-deep waters. Getting to these reservoirs involves ever-higher costs for drilling and well data sampling; it also involves increased investment in research and development (R&D) to successfully use cheaper and more efficient geophysical exploration methods. The recent high quality of geophysical data, particularly reflection seismic data, represented a breakthrough in reservoir modeling and characterization. However, the use of 3D and 4D seismic data has been a real challenge for geostatistical data integration methods.

Using seismic reflection data, stochastic seismic inversion methods are playing an important role in the characterization of oil and gas reservoirs, which can basically be divided into two groups: the first based on the linearized Bayesian approach to seismic inversion (Buland and Omre 2003; Tarantola 2005); the second based on geostatistical inversion methods that are essentially stochastic sequential simulations and optimization processes like genetic algorithms and simulated annealing (Bortolli et al. 1993; Haas and Dubrule 1994; Soares et al. 2007).

This book focuses on the geostatistical inversion methods, with Chap. 2 providing an overview of elementary geostatistics. Stochastic simulations and joint simulations are presented in Chap. 3, with the different versions used in the inversion approaches: joint simulation with joint distributions to deal with multivariate inversion and direct inversion and simulation with point distributions to access the data uncertainty.

In Chap. 4 we develop seismic inversion methods—acoustic, elastic and amplitude versus angle (AVA)—within the geostatistical framework. Chapter 5 encompasses the direct inversion of porosity, facies and rock physics models (RPM). Chapter 6 focuses on other methods of geophysical integration, such as joint electromagnetic and seismic inversion and the integration of seismic in history matching processes: that is, the integration of seismic and dynamic production data in numerical reservoir models.

This book is a natural extension and a summary of the Lisbon University Technical Institute's (IST—Instituto Superior Técnico) Master of Science program in Petroleum Engineering notes on reservoir characterization as well as those of the short courses given at other schools and oil companies. We are indebted to all the students whose critiques enriched this book.

The seismic inversion methods presented here were developed and implemented at the Centre for Modeling Petroleum Reservoirs research Centre, which is now the Petroleum Group of the Centre for Natural Resources and Environmental Research (CERENA—Centro de Recursos Naturais e Ambiente). We would like to express our thanks to Jean Paul Diet, who introduced us to the concept of geostatistical inversion and helped secure funding from CGG. We would also like to thank Thierry Colleau, who helped steer us in the right direction at the outset, and António Costa Silva, Luís Guerreiro and Carlos Maciel of Partex Oil and Gas, for their encouragement and support and for helping us find real applications for it in the Middle East.

Petrobras geophysicists Guenther Neto, Lucia Dillon and Evaldo Mundim brought new insights and their substantial experience to these methods. Last but not least, we would like to express our gratitude to the many researchers at the Centre for Modeling Petroleum Reservoir (CMRP) and our colleagues and friends, especially Maria João Pereira, Ana Horta, Rúben Nunes, Pedro Correia and Hugo Caetano, who have helped develop these methods over the past 15 years with ideas developed in the course of many projects, theses and publications.

Finally, a special acknowledge to Ângela Pereira, Catarina Marques and Pedro Pereira for the detailed revision of the manuscript and the valuable recommendations and Rúben Nunes for all the programming related with the DSS algorithm.

Lisboa, Portugal   Leonardo Azevedo
                                                             Amílcar Soares

Contents

List of Abbreviations and Mathematical Symbols

AVA	Amplitude versus Angle
AVO	Amplitude versus Offset
BHP	Bottom-hole pressure
$C_{i,j}(\mathbf{h})$	Cross covariance between variables $Z_i(.)$ and $Z_j(.)$ for distance $\mathbf{h}$
Cdf	Cumulative distribution function
CMP	Common midpoint
co-DSS	Direct Sequential co-Simulation
CSEM	Controlled source electromagnetic
C_0	Nugget effect
DSS	Direct sequential simulation
E{ }	Expected value
GEI	Global elastic inversion
GSI	Global stochastic inversion
$\mathbf{h}$	Distance vector
Ip	Acoustic impedance
Is	Elastic impedance
LFM	Low frequency model
MCMC	Markov chain Monte Carlo
MDS	Multidimensional scaling
OPR	Oil production rate
OK	Ordinary Kriging
Pdf	Probability distribution function
RC	Reflection coefficient
RF	Random function
RPM	Rock physics models
RV	Random variable
R(0)	Normal incidence
SGS	Sequential Gaussian simulation
SK	Simple kriging estimate
Sw	Water saturation
Vp	P-wave propagation velocity
Vs	S-wave propagation velocity
WPR	Water production rate
$\gamma(\mathbf{h})$	Variogram or semi-variogram for distance $\mathbf{h}$
$\gamma_{i,j}(\mathbf{h})$	Cross variogram between variables $Z_i(.)$ and $Z_j(.)$ for distance $\mathbf{h}$

σ_E^2	Kriging estimation variance
$\rho_{x,y}$	Pearson's correlation coefficient between variable x and y
$\rho(\mathbf{h})$	Correlogram as a function of $\mathbf{h}$
μ	Lagrange parameter

List of Figures

List of Tables

Introduction—Geostatistical Methods for Integrating Seismic Reflection Data into Subsurface Earth Models

1.1 Spatial Resolution Gap

The integration of seismic reflection data for subsurface modeling and characterization, while assessing the uncertainty of the subsurface property of interest, is becoming one of the most important challenges in reservoir characterization due to the increase in the quality of deep target geophysical—in particular seismic reflection data—information.

The direct use of seismic data as secondary data for reservoir characterization faces the problem of vertical spatial resolution gap between the low vertical resolution of seismic reflection data and the high vertical resolution of well-log data. Given these differences in spatial resolution, it is difficult to extract a relationship between, for example, seismic amplitudes and well-log porosity or acoustic impedance information from well-log data, to follow the traditional workflow of joint stochastic sequential simulation or stochastic sequential simulation with known means (Deutsch and Journel 1992).

More traditional workflows for integrating seismic reflection data for modeling the subsurface petrophysical properties (e.g. porosity, facies and saturation) begin by inferring the subsurface elastic properties (e.g. acoustic and/or elastic properties) from the available seismic reflection data gathered by any seismic inversion methodology. These seismic properties are represented in a more compatible resolution as the well-log data. From the resulting inverted elastic models the use of geostatistical techniques, such as stochastic sequential co-simulation (Chap. 3), allows the simulation of the petrophysical properties of interest by using the well-log data of the property to be modelled as the primary variable and the inverted elastic models as secondary variable. In this sequential approach, petrophysical modeling is performed in two independent steps (Fig. 1.1). The petrophysical property of interest is inferred from an inverted Earth model and is not a direct result of the inversion process; therefore, the resulting petrophysical model is not directly constrained by the available seismic reflection data.

Another approach is to interpret the seismic properties—acoustic impedance, for example—as a trend of the main petrophysical property: i.e. porosity. In these situations the stochastic simulation with local means (Chap. 3) can be applied. In this case, as the local trend is usually a smooth version of main property spatial dispersion, the eventual differences between spatial resolutions become less important. Even so, depending on the inversion method used to obtain the seismic property, the spatial resolution gap between seismic and well-log data, can still be an issue in these two approaches, i.e. joint stochastic sequential simulation and stochastic sequential simulation with known local trend.

© Springer International Publishing AG 2017
L. Azevedo and A. Soares, *Geostatistical Methods for Reservoir Geophysics*,
Advances in Oil and Gas Exploration & Production, DOI 10.1007/978-3-319-53201-1_1

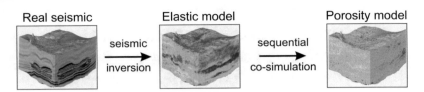

Fig. 1.1 Schematic representation of the traditional geo-modeling workflow to derive petrophysical models (e.g. porosity) from inverted elastic models

1.2 Seismic Inversion

This book focuses on different methods to integrate seismic data into subsurface Earth models: the seismic inversion. Seismic reflection data, or its amplitudes, is usually poorly related with subsurface petrophysical properties (such as facies, porosity and saturation). But, these properties often have a significant relationship with acoustic and/or elastic impedances. For example, the acoustic impedance, which is the product of rock density and velocity, usually has a good inverse relationship with porosity.

As the acoustic and/or elastic impedances are physically related with seismic amplitudes, $\mathbf{A}$, these can be approximated by a convolution of seismic reflectivity $\mathbf{r}$ (derived from acoustic/elastic impedances) with a known wavelet $\mathbf{w}$:

$$\mathbf{A} = \mathbf{r} * \mathbf{w}, \qquad (1.1)$$

where $\mathbf{A}$ is the recorded seismic amplitude obtained by the convolution of $\mathbf{r}$, the subsurface reflection coefficients, which are dependent on the elastic properties (P-wave and S-wave velocities and density) of the subsurface geology, with an estimated wavelet $\mathbf{w}$.

This physical relationship induced a new class of methods based on the inverse solution of the problem: one wants to know the model parameters (i.e. facies, porosity or related acoustic or elastic impedance), which derived reflectivity coefficients, convolved with a known wavelet, lead to the known solution: i.e. the recorded seismic amplitudes.

The theoretical solutions for seismic inversion are described in Tarantola (2005). The seismic inversion problem began to be approached with deterministic methodologies that are basically optimization procedures seeking the minimization of an objective function; normally the mismatch between the synthetic seismic reflection data obtained by perturbing an initial guess elastic model and the observed seismic reflection data.

Over recent years, seismic inversion has successfully been extended to a statistical framework for assessing the uncertainty of the inferred 3D subsurface elastic models, which is one of the main limitations of deterministic inverse procedures. Two different stochastic approaches to solve the seismic inversion ought to be noted. One approach of stochastic seismic inversion algorithms, the linearized Bayesian inversion (Buland and Omre 2003), are based on a particular solution of the inverse problem under the Bayesian framework. These methods assume the parameters and observations, as well as the data error, are multi-Gaussian distributed, which allows the model to be linearized (Chap. 4). Within this framework the posterior distribution is analytically obtained and is also multi-Gaussian. The second category comprises stochastic methodologies to tackle the seismic inversion as an optimization problem in an iterative and convergent process. This includes what are known as iterative geostatistical seismic inversion methods. They are based on stochastic and joint sequential simulations to generate possible solutions and an optimization procedure (e.g. genetic algorithm, simulated annealing) to guarantee convergence with the known seismic amplitudes. These geostatistical inverse methods are the focus of this book (Chap. 4), and are based on the family of direct stochastic sequential simulations and joint co-simulations with local and point distributions. As they do not imply any nonlinear transformation of parameters or observations, these methods have a high potential to accommodate accurate solutions for new challenges of different data integration, like the joint inversion of seismic reflection and electromagnetic data or the integration of production data into seismic inversion (Chap. 6).

Fundamental Geostatistical Tools for Data Integration

Geophysicists and other geoscientists are among the potential readers of this book. The purpose of this chapter is to summarize the basics of geostatistics, focusing on the set of methods used in stochastic simulations (Chap. 3) and seismic inversion (Chaps. 4 and 5). For those seeking more detailed information on geostatistics, we recommend Journel and Huijbreghts (1978), David (1977), Isaaks and Srivastava (1989), Goovaerts (1997) and Chilès and Delfiner (1999). As for the broad range of applications of geostatistics the following compilation of papers of different international geostatistics congresses are recommended: Verly et al. (1984), Armstrong et al. (1989), Soares (1993), Baafi and Schofield (1997), Kleingeld and Krige (2001), Leuangthong and Deutsch (2004), Ortiz and Emery (2008), Abrahamsen et al. (2012).

Geostatistics began playing a central role in the modeling and characterization workflows of hydrocarbon reservoir characterization in recent years (Dubrule 2003). By definition, geostatistics is a set of statistical tools that seek to describe the spatial and/or temporal distribution of a given property of interest, of which one only knows its value at sparse and discrete locations (Goovaerts 1997). Despite its potential, the use of geostatistical tools as part of the reservoir geo-modeling workflow is still traditionally restricted to the three-dimensional interpolation of the reservoirs' internal properties of interest (e.g. porosity, velocities) in between the sparse well locations.

The importance of geostatistical techniques has grown considerably largely as a result of their ability to integrate, within the same framework—the reservoir grid—geophysical and well-log data of a different nature and support. For example, the integration of seismic reflection data during the geo-modeling procedure allows for more detailed, heterogenic and reliable reservoir models when compared with those based exclusively on well-log data. This is due to the much higher spatial coverage the seismic data provides compared with the well data (Doyen 2007). Among the most well-known geostatistical algorithms are the Kriging methods (Sect. 2.3: Deutsch and Journel 1992) and conditional simulations [Chap. 3: e.g. sequential Gaussian simulation, direct sequential simulation; (Deutsch and Journel 1992; Gomez-Hernandez and Journel 1993; Verly 1993; Soares 2001)].

Deterministic models estimate the value of the property of interest, $z(x_0)^*$ at location x_0 by using a linear combination of the observed values, i.e. the experimental data. Within this framework, the inferred value is believed to correspond to the true value for that unknown location, $z(x_0) = z(x_0)^*$. The interpolated values are interpreted as having no associated error and the underlying assumption is that the physical system being modelled is fully known (Goovaerts 1997). By assuming no uncertainty in the inferred parameter, deterministic models are hardly suitable for describing

L. Azevedo and A. Soares, *Geostatistical Methods for Reservoir Geophysics*, Advances in Oil and Gas Exploration & Production, DOI 10.1007/978-3-319-53201-1_2

complex and heterogeneous systems like hydro-carbon reservoirs. In these environments, the lack of knowledge of the physical system being modelled is large, and uncertainty should be assessed during the modeling process (Caers 2011).

Unlike deterministic models, the use of a probabilistic framework reflects the lack of knowledge we have about the natural Earth system being modelled. At location x_0, the probabilistic framework provides a distribution of possible values for the property of interest along with its probability of occurrence, allowing the assessment of the spatial uncertainty of the property being modelled at a particular location of interest (Caers 2011). Finally, it is worth noting that any model resulting from a proba-bilistic approach is constrained by a set of assumptions about prior probability distributions that are estimated from available experimental data and the spatial continuity model imposed by, for instance, a variogram model or training image (Goovaerts 1997; Strebelle 2002).

This section introduces the main geostatistical stochastic sequential simulation approaches, due to their importance in assessing spatial uncer-tainty in recent modeling workflows and the lack of a real understanding of this family of algo-rithms within the oil and gas industry. Chapter 3 deals exclusively with stochastic sequential simulation algorithms. These algorithms are the basis of the geostatistical modeling techniques presented in Chaps. 4, 5 and 6.

2.1 Spatial Continuity Patterns Analysis and Modeling

Much of the success related with geostatistical models in Earth sciences relates to the ability to reproduce subsurface three-dimensional numeri-cal models with the relevant statistics of the variables retrieved from available experimental data. This reproduction is particularly effective for the spatial continuity and variability of the physical property under investigation.

The purpose of the geostatistical methodolo-gies for estimation, simulation and inversion introduced here is to generate numerical subsur-face models that reproduce the main statistics and spatial distribution as they are quantified (or estimated) as a result of available information and experimental data. Therefore, in this section we deal with the geostatistical tools that allow inference of the spatial continuity patterns of a natural resource for a given property measured at sparse locations within the study area.

Modeling the spatial behavior of a given property plays a key role in geostatistical methodologies, fulfilling two objectives: first, the characterization and quantification of the spatial pattern of a reservoir property, commonly des-ignated in geostatistics as spatial continuity analysis, i.e. the quantification of the spatial continuity for the property of interest and the way how it varies in different spatial directions; second it is also the basis for the spatial infer-ence/estimation, simulation and geostatistical inversion methodologies presented in the fol-lowing chapters.

2.1.1 Bi-point Statistics

Let us start with a simple example of an image, or two-dimensional model, with a biphasic phe-nomenon: a body X within a given area A, which is composed by X and its complementary X^c ($A = X \cup X^c$) in two distinct cases (Fig. 2.1), for which we intend to quantify the degree of spatial continuity.

To calculate the proportion of X in A, we can use a point that visits all possible locations within A and takes the value of '1' if it intersects body X, and the '0' if it intersects its complementary X^c. The ratio between the absolute frequency of intersections '1' and the total number of posi-tions within A is an estimator of the proportion of X in A.

Similarly, to measure the spatial continuity or dispersion of X, we can consider a circle with radius r and count the number of times it is

Fig. 2.1 Based on the circle as the basic structural element, the two images (**a**) and (**b**) have a spatial continuity of *X* as illustrated in (**c**)

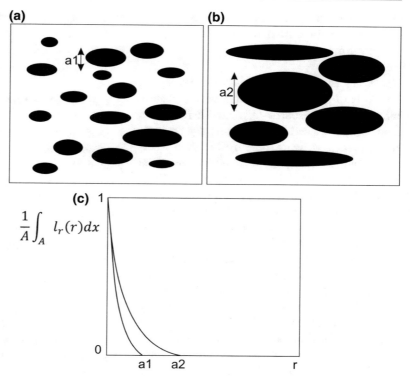

wholly contained in *X* while visiting all the locations within *A*. Increasing the radius *r* allows us to assess the continuity of *X* for both situations, assuming an indicator variable $I_r(x) = 1$ if the circle of radius *r*, centered in *x*, is wholly contained within *X*, and $I_r(x) = 0$ if the circle is not wholly contained in *X*. This relative measurement of the spatial continuity, which varies inversely with the size of *r*, is given by the following integral calculation (Eq. 2.1):

$$\frac{1}{A}\int_A l_r(x)dx. \tag{2.1}$$

In the case previously described, the circle of radius *r* is considered, in image processing and particularly for mathematical morphology, to be a structural element that allows the inference of the spatial continuity of *X* (Serra 1982). However, there are many other structural elements that may be used for the same purpose: a line segment *l* or the bi-point—pairs of points separated by a vector distance *h*—that are richer than the circle in a

morphological point of view. These tools, when compared with the circle as structural element, allow the measurement of other parameters, such as the anisotropy degree of *X*: i.e. the way the continuity of *X* varies in different spatial directions.

Although less rich than the line segment, the bi-point acting as structural element for measuring the spatial continuity of a natural resource is the privileged structural element in geostatistics. Notice that, unlike the line segment, the bi-point does not include the notion of connected sets (two points may simultaneously belong to *X* while not being connected). However, the knowledge we have from a given resource is not normally acquired from a two-dimensional representation, as in Fig. 2.1. In practice, we normally have access to a sparse limited discrete group of samples located within the study area (e.g. well data, core samples, soil samples). For this reason, the inference of the spatial continuity of a given property is frequently performed by returning to the bi-point as structural element (Sect. 2.1.4).

2.1.2 Complex Morphologic Patterns: Auxiliary and Reference Images

There are cases in which the connectivity of bodies is of outmost importance for the characterization of a resource (e.g. meandering sand channels in braided river sedimentary environments associated with some hydrocarbon reservoirs). In these cases, given the limitations of the bi-point in characterizing the connectivity of two distinct bodies (see above), we may have to return to proxy images of co-variables for the successful reproduction of these complex spatial patterns. The spatial distribution of these complex structures may be inferred from 2D and/or 3D images, conceptual models or interpreted from available geophysical data acquired from the hydrocarbon reservoir of interest, for example. Although they are auxiliary variables, which may be directly or indirectly related to the properties being studied, that information (e.g. seismic amplitudes, resistivity) may be used as auxiliary variables for joint simulation or as target images for geostatistical inversion, which then allows the reproduction of those complex spatial patterns within the geo-modeling workflow.

When geophysical data is scarce or unavailable, for example during early exploratory phases, another reliable alternative is to use feasible geological models of a given sedimentary environment. These representations of reality are built by gathering all the information about the system being studied (e.g. information from analogous and neighboring fields, expert opinion from geologists), and are often referred to as reference images. These reference images are then used to quantify the continuity and connectivity of the features from multi-point structural elements (Strebelle 2002; Arpat and Caers 2007; Mariethoz et al. 2010; Renard and Allard 2013; Mariethoz and Caers 2014), as in the example of the line segment.

Although these multi-point statistics methodologies are not the focus of this book, it is important to stress their use in complex geological environments. The choice of multi-point statistics as tools to quantify the spatial continuity patterns of a given property should be exclusively directed by the trust and knowledge of the degree of similitude between reality and the reference image.

2.1.3 Spatial Random Fields

A random variable (RV) is defined as one that can assume all the values contained within a probability distribution function. It can be continuous if the possible range of outcomes is continuous, or discrete if the outcomes are finite and without any specific order. By using the concept of RVs, the value of a property (e.g. porosity, acoustic impedance) at a given location within a study area (e.g. a reservoir grid) is interpreted as a single realization, $z(x_1)$ of the RV $Z(x_1)$. The group of these dependent RVs, located for example along a reservoir grid, is defined as a random field (RF) (Ventsel 1973).

To properly model a stochastic process there is no need to explicitly characterize the entire number of associated RVs and their corresponding multivariate distributions. Instead, and under some a priori assumptions, all that is required to spatially characterize a given property is to describe a certain number of parameters, such as the mean (Eq. 2.2) and variance (Eq. 2.3; Isaaks and Srivastava 1989):

$$E\{Z(x_i)\} = m(x_i) = \int_{-\infty}^{+\infty} z \, dF_{x_i}(z), \quad (2.2)$$

$$var\{Z(x_i)\} = \int_{-\infty}^{+\infty} [z - m(x_i)]^2 dF_{x_i}(z), \quad (2.3)$$

where $F_{x_i}(z)$ is the probability distribution function of the RV $Z(x_i)$.

If we consider two RVs, such as $Z(x_1)$ and $Z(x_2)$, the covariance between both variables is given by:

$$C(Z(x_1), Z(x_2)) = E\{Z(x_1)Z(x_2)\} \\ - m(x_1)m(x_2), \quad (2.4)$$

with

$$E\{Z(x_1), Z(x_2)\} = \int_{-\infty}^{+\infty} \int_{-\infty}^{+\infty} xy d^2 F_{x_1,x_2}(x,y),$$

$$(2.5)$$

where $F_{x_1,x_2}(x,y)$ is the bivariate probability distribution function

$$F_{x_1,x_2}(x,y) = prob\{Z(x_1) < x \text{ and } Z(x_2) < y\}.$$

$$(2.6)$$

The way the two RVs are spatially correlated is frequently described by a variogram model. The variogram between two RVs can be expressed as:

$$\gamma(Z(x_1), Z(x_2)) = E\{[Z(x_1) - Z(x_2)]^2\}. \quad (2.7)$$

Under the spatial RF assumption, the available experimental data is interpreted as being a single realization, $z(x_i)$, $i = 1, \ldots, N$ (with N equal to the total number of samples of the available experimental dataset), of a random function that comprises a set of spatially-correlated RVs. Therefore, by definition it is impossible to sample more than a single realization, $z(x_1)$, for a given RF. Even if the same location is sampled twice, each sample set will correspond to two different realizations of $z(x_i)$ in the same RF (Goovaerts 1997; Soares 2006).

However, a single realization of a random function is not enough to completely describe its statistical moments. The inference of these moments is only possible if we are somehow able to repeatedly sample a given location within the study area. The first and second statistical moments of a given RF can only be calculated by assuming different levels of stationarity within a specified study area: the stationarity of the mean and the stationarity of the spatial covariance (Goovaerts 1997; Soares 2006).

Within a pre-defined area of interest, A, the decision about the stationarity refers to a constant mean and spatial continuity pattern, as estimated from the available experimental data. By considering stationarity we can assume all RVs have the same mean within a limited area. With this assumption, the mean is not dependent on the

location, x_0, since it remains constant for the entire field. In this framework, the mean can then be estimated as the arithmetic mean of all the realizations of RF (i.e. the experimental dataset) composed by N samples, $Z(x_\alpha)$, $\alpha = 1, \ldots, N$:

$$m = \frac{1}{N} \sum_{i=1}^{N} Z(x_\alpha). \quad (2.8)$$

The second order of stationarity is defined if the correlation between two RVs depends exclusively on the distance between the two variables—the vector $\boldsymbol{h}$—and not on the specific location, x_0, of each variable. For example, the variogram between $Z(x_1)$ and $Z(x_2)$:

$$\gamma(Z(x_1), Z(x_2)) = \gamma(Z(x_1), Z(x_{1+\boldsymbol{h}})) = \gamma(\boldsymbol{h}).$$

$$(2.9)$$

By definition, the decision on stationarity can never be proved or refuted since we only know a single realization of the random function. Note that the stationarity is a property of the random function, or geostatistical model, needed to spatially infer the value of a given property far from the location of experimental data. This decision does not assume that the Earth's physical system we are trying to model is itself stationary. We should also test the available experimental data for its homogeneity across the entire study area. If this hypothesis cannot be assumed for the entire study area, then we may have to divide the field in smaller areas in which the decision about stationarity is more suitable (Goovaerts 1997).

2.1.4 Variograms and Spatial Covariances

Given a quantitative property, $Z(x)$, the diagrams representing the pairs of points, $Z(x)$, versus $Z(x + \boldsymbol{h})$ calculated for different values of $\boldsymbol{h}$ are the statistics parameters that contain more, and richer, information about the spatial continuity of $Z(x)$.

Figure 2.2 shows an example of well-log data with samples located along the well path. For each well we sample each pair of points with distance $\boldsymbol{h}$, $Z(x)$ and $Z(x + \boldsymbol{h})$. Figure 2.3 shows the cross-plots between pairs of points $Z(x)$ and $Z(x + \boldsymbol{h})$ for different values of $\boldsymbol{h}$ in the vertical direction: $h = 1, 2, 3, 10$. For $h = 1$ we may infer a good linear correlation between the values of samples $Z(x)$. This means there is a good correlation between the values of samples located in x and the values of the samples located immediately below. As soon as the values of h increase, the clouds of points start to scatter and the spatial correlation of the samples decreases. We may interpret from Fig. 2.3 that there is no correlation between samples separated by a distance of $h = 10$.

A group of diagrams constructed from different steps, $\boldsymbol{h}$, comprises almost all the information related with the degree of dispersion/ continuity for the variable, $Z(x)$, at that well location that we may retrieve from bi-point statistics. However, for a better interpretation and further use one must synthesize the bi-plots shown in Fig. 2.3 into a single tool. Summarizing the dispersion between pairs of points allows for a better visualization of the behavior of the property with increasing $\boldsymbol{h}$. One way, for example, would be to represent the correlation coefficients (Pearson's correlation) in function of $\boldsymbol{h}$,

resulting in what is commonly called a correlogram (Fig. 2.4).

In addition to the correlogram (Fig. 2.4), there are other measurements that synthesize the dispersion of different clouds of point $(Z(x), Z(x + \boldsymbol{h}))$, and which may result in a series of statistics quantifying the continuity of $Z(x)$. For example, each cross-plot from may be summarized by the mean of the least squares between $Z(x)$ and $Z(x + \boldsymbol{h})$, which is commonly called a variogram (or semi-variogram: Eq. 2.10):

$$\gamma(\mathbf{h}) = \frac{1}{2N(\mathbf{h})} \sum_{\alpha=1}^{N(\mathbf{h})} [Z(x_\alpha) - Z(x_\alpha + \mathbf{h})]^2, \quad (2.10)$$

where $N(\boldsymbol{h})$ is the number of pairs of points for each value of $\boldsymbol{h}$.

Note that all these statistics represent the spatial continuity of the variable $Z(x)$ in a given experimental location. In other words, the spatial continuity and its representativeness are limited to the region around the well. If instead of a single well we simultaneously consider all the wells with the same direction located within a reservoir and intersecting different geological layers, the resulting variogram (Eq. 2.10), or

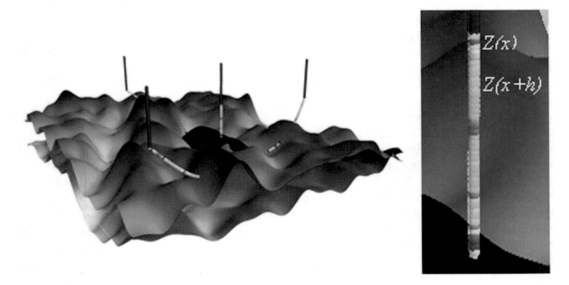

Fig. 2.2 *Left* Spatial representation of four wells and samples of a given subsurface property of interest measured along the well path. *Right* Detail of a well with the samples measured vertically along the well path

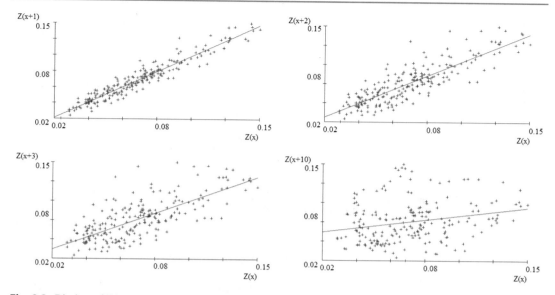

Fig. 2.3   Bi-plots of Z(x) versus Z(x + **h**) for different distances (values of **h**) in the vertical direction: $h = 1, 2, 3$ e 10

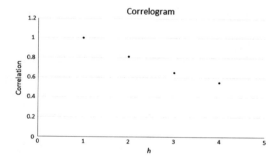

Fig. 2.4   Example of a correlogram for **h** between 1 and 4

correlogram, would represent the space covered by the group of available wells.

Let us now consider an example of a sedimentary environment related with large sinuous channels. This kind of geological setting is normally described as anisotropic, i.e. the spatial continuity/variability is different depending on the direction of space. For example, petrophysical properties, such as porosity and permeability, are frequently more continuous and consequently less variable along a given geological formation and are more variable between formations. The same exercise can be carried out using the pairs of values (Z(x), Z(x + **h**)) for different directions

in space. This exercise allows the assessment of spatial continuity for the property Z(x) along the entire domain of spatial analysis.

A different measurement of spatial continuity is given by the average of the product $Z(x)Z(x + h)$, for a given distance **h**. This results in a non-centered covariance estimate (Eq. 2.11):

$$C(\mathbf{h}) = \frac{1}{N(\mathbf{h})} \sum_{\alpha=1}^{N(\mathbf{h})} [Z(x_\alpha) \cdot Z(x_\alpha + \mathbf{h})]. \quad (2.11)$$

This covariance estimator may be centered by:

$$C(\mathbf{h}) = \frac{1}{N(\mathbf{h})} \sum_{\alpha=1}^{N(\mathbf{h})} [Z(x_\alpha) \cdot Z(x_\alpha + \mathbf{h})] - m(x_\alpha), m(x_\alpha + \mathbf{h}), \quad (2.12)$$

given $m(x_\alpha) = \frac{1}{N(h)} \sum_{\alpha=1}^{N(h)} Z(x_\alpha)$ and $m(x_\alpha + h) = \frac{1}{N(h)} \sum_{\alpha=1}^{N(h)} Z(x_\alpha + h)$ as the arithmetical averages for all the points at locations x_α and $x_\alpha + h$, $\alpha = 1, \ldots, N(h)$.

The covariance estimator (Eq. 2.12) may be expressed in terms of a correlogram (or normalized covariance):

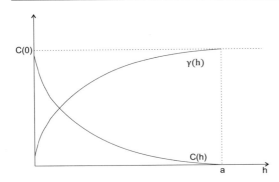

Fig. 2.5 Relationship between covariance, $C(\boldsymbol{h})$, and variogram, $\gamma(\boldsymbol{h})$, functions with the increment of step $\boldsymbol{h}$

$$\rho(\boldsymbol{h})\frac{C(\boldsymbol{h})}{\sqrt{\sigma^2_{(x_\alpha)} \cdot \sigma^2_{(x_\alpha+\boldsymbol{h})}}}, \qquad (2.13)$$

where

$$\sigma^2_{(x_\alpha)} = \frac{1}{N(\mathbf{h})}\sum_{\alpha=1}^{N(\mathbf{h})}[Z(x_\alpha) - m(x_\alpha)]^2,$$

and

$$\sigma^2_{(x_\alpha+\boldsymbol{h})} = \frac{1}{N(\mathbf{h})}\sum_{\alpha=1}^{N(\mathbf{h})}[Z(x_\alpha+\boldsymbol{h}) - m(x_\alpha+\boldsymbol{h})]^2.$$

By assuming the stationarity of increments $\boldsymbol{h}$ (Sect. 2.1.3), the mean of the least squares and the mean of the products are estimates of the second moments: the variogram (Eq. 2.14) and the centred covariance (Eq. 2.15):

$$\gamma(\boldsymbol{h}) = \frac{1}{2}E\{[Z(x) - Z(x+\boldsymbol{h})]^2\}, \qquad (2.14)$$

$$\begin{aligned}C(\boldsymbol{h}) &= E\{Z(x)Z(x+\boldsymbol{h})\}\\ &\quad - E\{Z(x)\}E\{Z(x+\boldsymbol{h})\},\end{aligned} \qquad (2.15)$$

we reach the relationship between a variogram and the covariance (Eq. 2.16):

$$\gamma(\boldsymbol{h}) = C(0) - C(\boldsymbol{h}). \qquad (2.16)$$

We then may express the variogram in terms of a correlogram (Eq. 2.17):

$$\rho(\boldsymbol{h}) = \frac{C(\boldsymbol{h})}{C(0)}. \qquad (2.17)$$

The relationship between the variogram and the covariance functions (Eq. 2.16) is synthesized in Fig. 2.5.

2.1.5 Spatial Representativeness of the Variogram

As noted in Sect. 2.1.3. above, it is important to relate the stationary assumption of the probabilistic model to the notion of the representativeness and homogeneity of the experimental samples, which are the basis for calculating the stationary statistics (mean and variance).

For the first statistical moment, the mean of the N samples within an area A, $m_z = \frac{1}{N}\sum_{\alpha=1}^{N} z(x_\alpha)$ is an estimator of the expected value of the random function $Z(x) - E\{Z(x)\}$, under the assumption of stationary about the mean.

Given that we only have access to a single realization of the random function $Z(x)$, i.e. the set of values $z(x_\alpha)$ in A, we ensure that m_z is a good estimate of the spatial integral (Eq. 2.18):

$$m_z = \frac{1}{A}\int_A z(x)dx. \qquad (2.18)$$

Assuming a stationarity mean for the random function $Z(x)$ is equivalent to considering the set of available experimental data—the only known realization of the random function $Z(x)$—as homogenous and representative for the entire area A. The same principle can easily be extrapolated for the second statistical moments: the variograms (Eq. 2.14) and covariance (Eq. 2.15).

Let us consider a given area, A, as a closed body and a distance vector, $\boldsymbol{h}$, as schematically represented in Fig. 2.6. A new area, $A_{+\boldsymbol{h}}$, can then be defined by the samples $x_{+\boldsymbol{h}}$ with distance $+\boldsymbol{h}$ from the samples x in A (Eq. 2.19):

$$A_{+\boldsymbol{h}} : \{x+\boldsymbol{h}|x \in A\} \text{ or } \{x|x - \boldsymbol{h} \in A\}. \qquad (2.19)$$

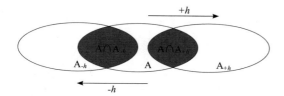

Fig. 2.6 Schematic representation of the spatial representativeness for covariance and variance estimates

In the same way we may mathematically define A_{-h} (Eq. 2.20):

$$A_{-h} : \{x - h | x \in A\} \text{ or } \{x | x + h \in A\}. \quad (2.20)$$

The variogram is by definition the variance of the deviations $(Z(x) - Z(x + h))$ when both samples x and $x + h$ belong to A. Hence, its representativeness refers to the union of both grey areas as shown in Fig. 2.6: $(A \cap A_{+h}) \cup (A \cap A_{-h})$.

In terms of spatial integrals of A, the variogram may be described as follows:

$$2\gamma(h) = \left[\frac{1}{A \cap A_{+h}} \int\limits_{A \cap A_{+h}} [Z(x) - Z(x+h)]^2 dx \right.$$

$$\left. + \frac{1}{A \cap A_{-h}} \int\limits_{A \cap A_{-h}} [Z(x) - Z(x-h)]^2 dx \right]. \quad (2.21)$$

When the distance vector h is small—for example smaller than half the dimension of A—then the representativeness of the variogram is similar to A:

$$(A \cap A_{+h}) \cup (A \cap A_{-h}) \approx A.$$

Thus, when computing the variogram estimate we must take into account that its representativeness in A is given by the dimension of the distance vector h. For practical applications, the representativeness of the variogram ($\gamma(h)$) should be questioned for distances, h, larger than half the size of A along the direction of h.

Besides the spatial representativeness, each value of the variogram must be related to a homogeneous cloud of pairs of points in the bi-plot $(Z(x), Z(x+h))$ for each step h. For example, one single anomalous value $Z(x_i)$ may result, along with its neighbor samples $Z(x_i + h)$, in large values of $[Z(x_i) - Z(x_i + h)]^2$, and consequently a large mean $\gamma(h)$ translated in practice as a weak spatial correlation for that given value of h. We may consider this sample anomalous and as having a restrict representativeness. If the computed experimental variogram, without taking into account the anomalous value $Z(x_i)$, has a more regular behavior (in a way that it is more similar to the rest of the values of $\gamma(h)$), it may and should be adopted as representative of the remaining samples for the whole area. Note that in these cases, during the process of local estimation the areas surrounding $Z(x_i)$ should be considered with special care to ensure the area of influence of the sample does not have a large impact on the estimate. By definition, $\gamma(h)$ does not translate the behaviour of the large spatial variability between $Z(x_i)$ and the neighbor samples.

2.1.6 Spatial Continuity for Multivariate Systems

Consider those cases in which, at a given spatial location for a single sample, we measure more than one attribute, $Z_1(x_i)$, $Z_2(x_i)$, ... $Z_N(x_i)$: for example, P-wave velocity, S-wave velocity and density measured at the same locations along a well path.

The correlation between each pair of these attributes $Z_1(x)$, $Z_2(x)$, ... $Z_N(x)$ may be measured through the correlation coefficient for the set of N samples (Eq. 2.22):

$$\rho(Z_1, Z_2) = \frac{1}{N\sigma_1\sigma_2} \sum_{i=1}^{N} [(Z_1(x_i) - m_1)(Z_2(x_i) - m_2)], \quad (2.22)$$

where m_1, m_2, σ_1^2 e σ_2^2 are the mean and variance of $Z_1(x)$ and $Z_2(x)$ respectively.

We can now generalize the correlation between the different variables and calculate the correlation of variable $Z_1(x)$ located in x and the variable $Z_2(x + h)$ located in $x + h$. The spatial dependency between each pair of variables with distance h may be characterized by cross-variograms, cross-covariance and cross-correlation (Goovaerts 1997).

Assessing the spatial dependency between variables is important as we often have to use an auxiliary variable (frequently more abundant) to estimate a primary variable (less abundant) assuming a spatial correlation between both. This is frequently the case when estimating porosity models with known values at the well locations from models of acoustic impedance retrieved, for example, from seismic inversion in the entire area A.

The random function model used so far (Sect. 2.1.3) may be generalized for multivariate cases. The set of RVs defining N_v random function, $I = 1, ..., N_v$, can also be designated as multivariate random function:

$$Z_i(x), \quad i = 1, \ldots, N_v, \quad \forall_{x \in A}. \quad (2.23)$$

The joint distribution of two variables $Z_i(x)$ and $Z_j(x)$ depends on the distance vector h (Eq. 2.24):

$$F_{ij}(h, z_i, z_j) = prob\{Z_i(x) \leq z_i, Z_j(x+h) \leq z_j\}, \quad \forall_{ij}. \quad (2.24)$$

The spatial dependency between two variables $Z_i(x)$ and $Z_j(x)$ may be measured by the cross-covariance function:

$$C_{ij}(h) = E\{[Z_i(x) - m_i] \cdot [Z_j(x+h) - m_j]\}, \quad \forall_{ij}, \quad (2.25)$$

or by the cross-variogram:

$$\gamma_{ij}(h) = \frac{1}{2}E\{[Z_i(x) - Z_i(x+h)] \cdot [Z_j(x) - Z_j(x+h)]\}, \quad \forall_{ij}. \quad (2.26)$$

Note that $\gamma_{ij}(h) = \gamma_{ji}(h)$, but $C_{ij}(h)$ may not equal $C(h)$, meaning the function is not symmetrical with h.

The relationship between cross-variogram and cross-covariance may be described as:

$$\gamma_{ij}(h) = C_{ij}(0) - \frac{1}{2}[C_{ij}(h) + C_{ij}(-h)]. \quad (2.27)$$

If the cross-covariance in Eq. 2.25 is rewritten in terms of the sum between two terms dependent on h (Eq. 2.27), it can easily be understood that the cross-variogram (Eq. 2.27) comprises only the first term. This is the reason for the symmetry around h:

$$C_{ij}(h) = \frac{1}{2}[C_{ij}(h) + C_{ij}(-h)] + \frac{1}{2}[C_{ij}(h) - C_{ij}(-h)]. \quad (2.28)$$

In practice, the asymmetry component of the cross-covariance is usually ignored for two main reasons (Journel and Huijbreghts 1978):

- The amount of available experimental data rarely allows comprehension and consequent validation across the whole study area of the physical phenomenon that results in the asymmetry in the cross-covariance;
- Modeling the asymmetric cross-covariance is extremely complex.

For this reason, the geostatistical tools that quantify the spatial continuity of a multivariate system are frequently the cross-variograms and the symmetrical cross-covariance: $C_{ij}(h) = C_{ji}(h)$, or the mean of $C_{ij}(h)$ and $C_{ji}(h)$.

Finally, the cross-correlogram may be synthetized by (Eq. 2.29):

$$\rho_{ij}(h) \frac{C_{ij}(h)}{\sqrt{C_{ii}(0) \cdot C_{jj}(0)}} \in [-1, 1]. \quad (2.29)$$

2.1.7   Variogram Modeling Workflow

Modeling by a Mean Representative Function

Figure 2.7a represents a given reality, i.e. a model with all the values of $Z(x)$ within an area A. From that reference data $Z(x)$, a limited set of experimental data was randomly sampled (black circles on Fig. 2.7a) and an experimental variogram calculated (Fig. 2.7b). As a reference dataset from which reality is known, we may also calculate the experimental variogram for all points within the study area, the entire $\gamma(h)$ (Fig. 2.7c).

The example illustrated in Fig. 2.7 synthesizes the main objective during the variogram modeling stage: the estimation of the real variogram using a discrete and limited set of samples $z(x)$ and the corresponding experimental variogram.

Once the values of the variograms for different distances, h, are calculated for a given area A, it is necessary to model them using a function describing the spatial behavior of the property of interest for the entire study area. In practice, we adjust a smooth function of a reduced number of parameters that describe the spatial continuity of $Z(x)$.

This step is of utmost importance within the geostatistical framework, since it allows the synthesis of the structural characteristics of the spatial phenomena, e.g. degree of dispersion/continuity and anisotropies, into a single and coherent variogram model.

It is also common practice to adjust a model to the experimental variogram by conditioning it from expert knowledge about the phenomena being modelled.

Positive Definite Models

From the many functions that may be used to interpolate the points of an experimental variogram we need to constrain our options to those allowing stable solutions when calculating linear estimates (Sect. 2.2). To meet this condition, the variogram and covariance must be positive definite. The necessary condition for the positive definite of a covariance matrix is:

$$\sum_i \sum_j \lambda_i \lambda_j C(i,j) \geq 0. \qquad (2.30)$$

Any linear combination of covariance between pairs of points within an area A is always positive definite. If we consider a given

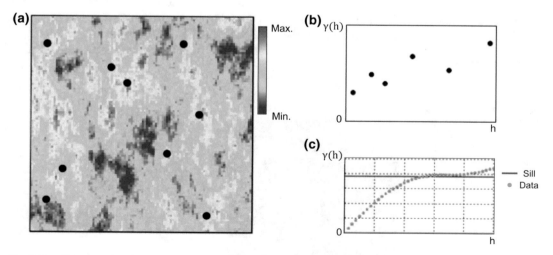

Fig. 2.7 **a** Known given data from an area A and set of samples retrieved from this data (*black filled circles*). **b** Experimental variogram computed from the set of available samples. **c** Experimental variogram computed from the known given data (**a**)

variable $Z(x_0)$ resulting from a linear combination of RVs $Z(x_1)$, $Z(x_2)$, ... $Z(x_N)$ (Eq. 2.31):

$$Z(x_0) = \sum_i \lambda_i Z(x_i), \qquad (2.31)$$

then a definite positive covariance ensures that the variance of $Z(x_0)$ is always positive:

$$var\{Z(x_0)\} = E\left\{ \sum_i \sum_j \lambda_i \lambda_j Z(x_i) Z(x_j) - m^2 \right\}$$

$$= \sum_i \sum_j \lambda_i \lambda_j E\{Z(x_i) Z(x_j) - m^2\}$$

$$= \sum_i \sum_j \lambda_i \lambda_j C(i,j) \geq 0. \qquad (2.32)$$

By replacing Eq. 2.16 in Eq. 2.31 the variance may be written in function of the variogram:

$$var\{Z(x_0)\} = C(0) \sum_i \sum_j \lambda_i \lambda_j$$

$$- \sum_i \sum_j \lambda_i \lambda_j \gamma(i,j) \geq 0. \quad (2.33)$$

In the cases in which $C(0)$ does not exist (i.e. non-stationary random functions) the variance $Z(x_0)$ exists if $\sum_i \lambda_i = 0$. Thus, the necessary condition of positive variance is ensured if $\sum_i \sum_j \lambda_i \lambda_j \gamma(i,j) \leq 0$ conditioned to the sum of the weights being zero.

2.1.8 Theoretical Variogram Models

The positive definite condition limits the number of models that can be used for interpolating experimental variograms. In practice, within a geostatistical framework a limited range of positive definite interpolating functions are used. The following models are presented: spherical, exponential, Gaussian and power.

Spherical Model

The spherical model is one of the most common in geostatistics and it is a function of two parameters (Eq. 2.34): the sill (C), upper limit to which the values of the variogram tend when h is increased; and the range, a, distance from where the values of $\gamma(h)$ stop increasing and are approximately equal to the sill, normally the total variance of the experimental data $Z(x)$. The range of a variogram measures the distance from where the data $Z(x)$ is no longer correlated:

$$\gamma(h) = \begin{cases} C\left[1.5\frac{h}{a} - 0.5\left(\frac{h}{a}\right)^3 \right] & for \quad h \leq a \\ C & for \quad h > a. \end{cases}$$

$$(2.34)$$

Figure 2.8a shows a map with the spatial distribution of a variable $Z(x)$ modelled with a spherical model (Fig. 2.8b) in which the amplitude is one-third of the dimension map length.

Fig. 2.8 a Map with the spatial distribution of the variable Z(x) according to **b** a spherical model with amplitude equal to 1/3 L

(a)

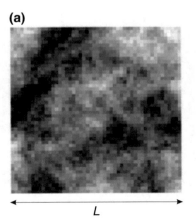

L

(b)

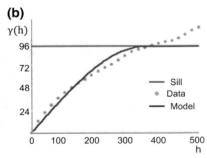

Fig. 2.9 **a** Map with the spatial distribution of the variable Z(x) according to **b** an exponential model with amplitude equal to 0.3 L

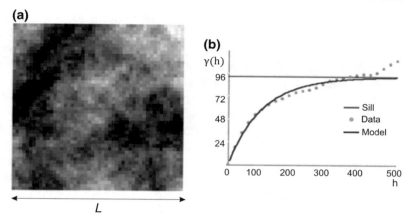

Exponential Model

The exponential model is a function of the same parameters of the spherical model—sill and range (Eq. 2.35)—while the variogram tends asymptotic to the sill value:

$$\gamma(\boldsymbol{h}) = C\left[1 - \exp\left(-\frac{3\boldsymbol{h}}{a}\right)\right]. \tag{2.35}$$

In this model, the range value is the distance in which the model reaches 95% of the sill: $\gamma(\boldsymbol{h}) = 0.95\,C$.

Figure 2.9a shows the spatial distribution of a variable modelled with an exponential variogram (Fig. 2.9b) with range equal to 0.3 L of the map length.

Comparing both Figs. 2.8 and 2.9, besides the rapid growth of the spherical model near the origin, it shows structures with larger spatial continuities resulting from larger spatial correlation for larger distances *h*.

Gaussian Model

The two variogram models previously presented —the spherical and exponential—have a relatively fast increase near the origin, translating into a typical behavior of irregular natural phenomena. Other phenomena, more regular and continuous, are translated by a slow increase of the variogram values near the origin, for example a parabolic behavior. This is the case of Gaussian models (Eq. 2.36):

$$\gamma(\boldsymbol{h}) = C\left[1 - \exp\left(\frac{-3\boldsymbol{h}^2}{a^2}\right)\right]. \tag{2.36}$$

Fig. 2.10 **a** Map with the spatial distribution of the variable Z(x) according to **b** a Gaussian model with amplitude equal to 0.2 L

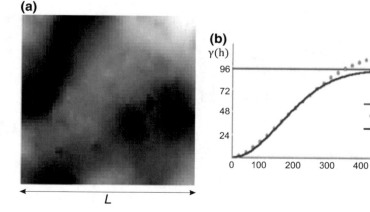

As in the exponential model, range, a, is the distance from where the model reaches 95% of the sill: $\gamma(\boldsymbol{h}) = 0.95\,C$.

Figure 2.10 shows the behavior of a Gaussian variable model in which the range is 0.2 L of the length of the map. Note the much larger and smooth spatial continuity compared with those obtained from the spherical and exponential models (Figs. 2.8 and 2.9).

Power Models

So far all the variogram models described have a sill as upper limit for where the variogram values of $\gamma(\boldsymbol{h})$ tend when $\boldsymbol{h}$ increases infinitely. These models are adequate for transition phenomena characterized by a distance—the range—from where the spatial correlation between samples no longer exists. In these transitional phenomena there is always a relationship between covariance and variogram as described in Eq. 2.16.

However, there are other natural phenomena in which the growth of $\gamma(\boldsymbol{h})$ is continuous with h and does not tend to the sill. These are non-stationary phenomena in which there is no finite variance or notion of covariance—the variance grows with the dimension of the dispersion field $Z(x)$.

The most common variogram model applied in these situations is the power model (Eq. 2.37):

$$\gamma(\boldsymbol{h}) = C\boldsymbol{h}^{\alpha}, \qquad (2.37)$$

with α between 0 and 2. Depending on α, the variogram may be linear ($\alpha = 1$), logarithmic ($0 < \alpha < 1$) or parabolic ($1 < \alpha < 2$) (Fig. 2.11).

2.1.9 Linear Combinations of Variogram Models: Imbricated Structures

For most natural phenomena, spatial continuity patterns are rarely simple and modelled by a single variogram model. Normally different structures coexist simultaneously with distinct spatial continuities and distinct characteristics. A simple example is the one illustrated by Fig. 2.12, which shows a binary process with structures at two different scales: the first structure is composed of bodies with average size a_1; the grouping of these structures results in bodies with dimension a_2 (second structure).

In these cases, the experimental variogram reveals the simultaneous effect of the imbricated structures through discontinuities in the growing behavior of the variogram values with the distance $\boldsymbol{h}$.

For comparing the effect of modeling a property with different variogram models, Fig. 2.13 shows four 2D models (with size 1000×1000 m) with distinct spatial variogram models, respectively: (a) a spherical model with range equal to 150 m; (b) a spherical model with range equal to 750 m. Figure 2.13c, d show two imbricated structures with different contributions:

- Figure 2.13c has a spatial distribution modelled by a weighted mean of 70% for the first structure (a = 150 m) and 30% for the second

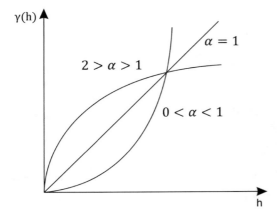

Fig. 2.11 Schematic representation of power models

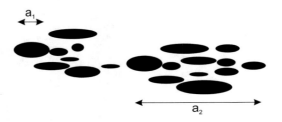

Fig. 2.12 Example of a binary process with structures at two different scales, a_1 and a_2

Fig. 2.13 Images with the same attribute modelled with different variogram models: **a** spherical model with zero nugget effect for a single structure with range 150 m; **b** spherical model with zero nugget effect for a single structure with range 750 m; **c** imbricated structure quantified by a weighted mean of 70% for a first structure with a = 150 m and 30% for the second structure with a = 750 m; **d** imbricated structure quantified by a weighted mean of 30% for a first structure with a = 150 m and 70% for the second structure with a = 750 m

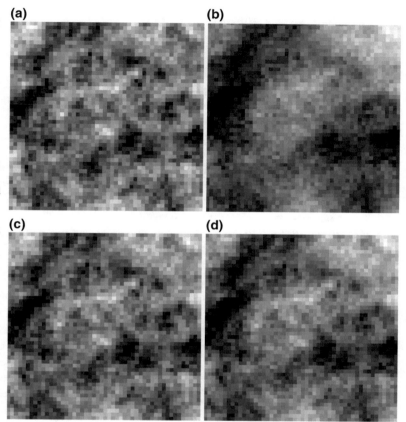

(a = 750 m). The greater influence of the smallest structure is clear. The resulting variogram model may be expressed by the following linear combination:

$$\gamma(\boldsymbol{h}) = 0.7\ Sph(a = 150\,m) + 0.3\ Sph(a = 750\,m).$$

- Figure 2.13d has a spatial distribution modeled by a weighted mean of 30% for the first structure and 70% for the second. In´ this image the greater influence of the largest scale structure is clear. The resulting variogram model may be synthetized by the following linear combination:

$$\gamma(\boldsymbol{h}) = 0.3\ Sph(a = 150\,m) + 0.7\ Sph(a = 750\,m).$$

The imbricated structures are modelled through a linear combination of the variogram models presented above (Fig. 2.14). They benefit from the propriety of the positive definite models:

in any linear combination of positive coefficients of models definite positive is positive. Any linear combination of variogram models (Eq. 2.38) is an imbricated group of variogram $\gamma_i(\boldsymbol{h})$ where the weights $C_i(0)$ are the sills—total variance—for each single structure. The sum of the different sills is equal to the global sill:

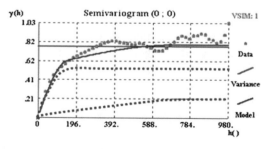

Fig. 2.14 Variogram model resulting from the sum of two distinct structures: $\gamma(\boldsymbol{h}) = 0.48$ Sph (a = 150 m) + 0.24 Sph (a = 750 m)

$$\gamma(\boldsymbol{h}) = \sum_i C_i(0)\gamma_i(\boldsymbol{h}). \qquad (2.38)$$

Nugget Effect

Theoretically, the value of the variogram is zero for $h = 0$ ($\gamma(\boldsymbol{h}) = 0$, for $h = 0$). In practice, between consecutive samples there is a minimum value of $\boldsymbol{h}$ for which the value of $\gamma(\boldsymbol{h})$ may be calculated. When this minimum value ($\gamma(h_{min})$) is high, it means there is a high variability in the natural phenomenon at the small-scale, i.e. for distances smaller than the distances between samples or observations: $\gamma(\boldsymbol{h})$ may not tend to zero while h tends to zero. In these cases, there is an inflexion or discontinuity in the growth of the variogram at a scale not sampled by the available experimental data, i.e. $h = 0$ and h_{min}. For these cases, the variogram is modelled by a constant, C_0, which is called the nugget effect (Eq. 2.39). The nugget effect is the first structure summed to the linear combination of the remainder of the structures:

$$\gamma(\boldsymbol{h}) = C_0 + \sum_i C_i(0)\gamma_i(\boldsymbol{h}), \qquad (2.39)$$

while the total variance is described by:

$$C(0) = C_0 + \sum_i C_i(0). \qquad (2.40)$$

The nugget effect summarizes the effect of two distinct parts of the total variability in the natural phenomenon being modelled: (i) the small-scale variability not comprised in the sampling grid; (ii) the variability at the scale support sample introduced by non-systematic sampling errors that add to the structure of the natural phenomenon random noise component.

Modeling the nugget effect by a constant different from zero translates the lack of knowledge about the system at the small-scale by increasing the uncertainty during the estimation procedure (Sect. 2.2 and Chap. 3). The nugget effect can be interpreted as the intersect value at the ordinate axis. It is usually inferred by the intersection of a line approximating the first

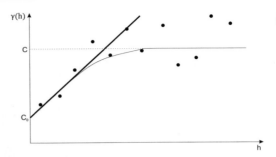

Fig. 2.15 Nugget effect inference based on the linear regression (*thick black line*) of the first points of $\gamma(\boldsymbol{h})$. Variogram model is represented by the thin black like

points of the variogram with the ordinate axis (Fig. 2.15).

Anisotropy Models

The spatial continuity of a natural resource frequently varies as a function of the direction of the space, e.g. greater continuity for porosity values along a channelized structure versus lower continuity across the channel direction. A given attribute of any natural resource (e.g. porosity in hydrocarbon reservoirs) has an isotropic spatial continuity if the variogram (or covariance) has the same behavior in all directions (that is, $\gamma(\boldsymbol{h})$ depends exclusively on the modulus of the distance vector $\boldsymbol{h}$). There are cases, however, in which the attribute being studied is more continuous along a preferential direction resulting in structural anisotropy. The latter may be seen as the variability of a given attribute depending on the directions of the space we consider when inferring the spatial behavior of a specific property.

Modeling anisotropic structures seeks to reduce structures of continuity depending on the direction of a single variogram model. The anisotropy is normally processed in terms of geometric transforms of the coordinate system in such a way as that the several variograms along different directions are equivalent to a single model, transforming them into isotropic structures.

Two of the most common anisotropy models are the geometric and the zonal.

Geometric Anisotropy

Geometric anisotropy is a model in which the spatial continuity, revealed by the variogram amplitudes, varies gradually from the direction of larger continuity/range through the direction of smallest amplitude, perpendicular to the first following the equation of an ellipse defined by those directions.

A geometric anisotropy is characterized by variograms with the same model and the same sill in all directions, but different ranges, while the minimum and maximum amplitudes are in perpendicular directions (Fig. 2.16).

Geometric anisotropy means the rose diagram that describes the ranges of the variogram along the different directions of the space may be modelled by an ellipse in 2D or an ellipsoid in 3D. In fact, when representing the different variogram ranges as a function of the angle of the variogram direction in a diagram, the ellipse is the geometric figure that best describes a geometric anisotropy.

An ellipse may be seen as a linear transformation of a circle (or sphere in three-dimensions), which corresponds to an isotropic condition: that is to say, the variogram range does not change with the variogram direction. A simple method for combining a group of variograms, with ranges ax, ay and az in the three directions of space, respectively, into a single model with range $a = 1$ is given by the following geometric transformation:

$$\gamma_{a=1}(\boldsymbol{h}) = \gamma_x(\boldsymbol{h}_x) \quad \text{with } h = h_x/a_x,$$
$$\gamma_{a=1}(\boldsymbol{h}) = \gamma_y(\boldsymbol{h}_y) \quad \text{with } h = h_y/a_y. \quad (2.41)$$
$$\gamma_{a=1}(\boldsymbol{h}) = \gamma_z(\boldsymbol{h}_z) \quad \text{with } h = h_z/a_z.$$

This corresponds to the normalization of the distances within the Cartesian space as follows:

$$h = \sqrt{\left(\frac{h_x}{a_x}\right)^2 + \left(\frac{h_y}{a_y}\right)^2 + \left(\frac{h_z}{a_z}\right)^2}. \quad (2.42)$$
$$\gamma_{a=1}(h) = \gamma_x(hx) \text{ if } h = h_x, /a_x$$

If we choose to transform the anisotropy into a reference variogram (for example, the variogram with the largest range instead of a variogram with range equal to 1), the normalized distance h can be described as:

$$h = \sqrt{h_x\left(\frac{a_x}{a_x}\right)^2 + h_y\left(\frac{a_y}{a_y}\right)^2 + hz\left(\frac{a_z}{a_z}\right)^2}, \quad (2.43)$$

where ax is the reference variogram range and $r_x = a_x/a_x$, $r_y = a_x/a_y$ e $r_z = a_x/a_z$ are the three anisotropic ranges in the three main ranges.

This methodology of transforming coordinates may be applied equally to non-stationary models that do not reach the sill.

If the direction of largest range does not match the axis of the reference coordinate system, it must be rotated so the axis xx' matches ax, yy' matches ay and zz' matches az before applying the geometry transform (Eq. 2.43).

Figure 2.17 shows a two-dimensional example in which the maximum range is observed in the direction of 45° and the smallest amplitude perpendicular to this (135°). Therefore, any vector h must be first rotated 45° (Eq. 2.44) before any normalization operation:

$$\begin{bmatrix} h_{x_{45°}} \\ h_{y_{45°}} \end{bmatrix} = \begin{bmatrix} \cos 45° & -\sin 45° \\ \sin 45° & \cos 45° \end{bmatrix} \cdot \begin{bmatrix} h_x \\ h_y \end{bmatrix}. \quad (2.44)$$

Zonal Anisotropy

Zonal anisotropy is common in stratified phenomena in which the spatial continuity along a stratum is in contrast with the variability between

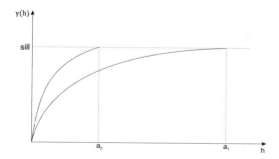

Fig. 2.16 Schematic representation of geometrical anisotropy: variograms with the same sill, but different ranges

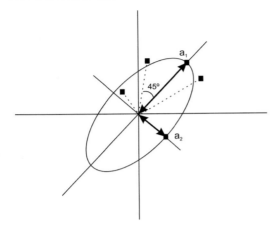

Fig. 2.17 Schematic representation of the rose diagram for the group of ranges and geometric anisotropic model

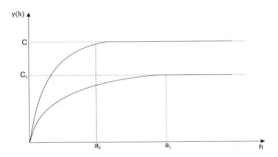

Fig. 2.18 Example of zonal anisotropy

a stratum and, consequently, the variogram along the stratum does not reach the sill of the variogram (Fig. 2.18).

The zonal anisotropy may be modelled by a linear combination of two structures:

$$\gamma(\boldsymbol{h}) = C_1\gamma_1(\boldsymbol{h}_1) + (C_2 - C_1)\gamma_2(\boldsymbol{h}_2), \quad (2.45)$$

where the second structure $\gamma_2(\boldsymbol{h}_2)$, here the vector $\boldsymbol{h}_{2,}$, is related only to the direction of larger variability—that is, between stratum. The example of Fig. 2.18 may be modelled by a first structure with a sill C_1 and a geometric anisotropy with an anisotropic ratio and a second structure with variance equal to $C_2 - C_1$ and range a_2 that only exists in direction $\boldsymbol{h}_2$:

$$\gamma(\boldsymbol{h}) = C_1\gamma_1(\boldsymbol{h}_1) + (C_2 - C_1)\gamma_2(\boldsymbol{h}_2),$$

$$h = \sqrt{\left(\frac{h_x}{h_x}\right)^2 + \left(\frac{h_z}{a_z}\right)^2}$$

$$and\ h_2 = \frac{h_z}{a_z}.$$

The two structures are defined as follows:

$\gamma_1(\boldsymbol{h})$—an isotropic model with a sill equal to C_1, and with ranges a_1 in direction h_1 and a_2 in direction h_2.

$\gamma_2(\boldsymbol{h})$—anisotropic model with a sill equal to $C_2 - C_1$, with range a_2 in direction h_2 and with 'infinite' range along h_1. Note that choosing a very large range in direction h_1 (for example, 10 times the dimension of the field) implies a small contribution of $\gamma_1(\boldsymbol{h})$ for values of h_1 near the dimension of the field.

Structural Transforms

There are cases in which the anisotropic phenomena do not regularly vary from the maximum to the minimum range: for example, when the spatial dispersion is conditioned by external factors to the genesis of this resource. Structural geology, such as folds and faults, are typical events in which these cases occur.

Each of these situations requires specific geometric transforms in order to achieve simple and generalized variogram models for the entire study area. For example, the transform associated with a folded geological formation, or dome, into a regular shape in order to better identify the spatial continuity between samples within a very non-isotropic structure (Fig. 2.19) (Mallet 2002, 2004). The conceptual basis of this transform is the following: the value of a sample in the thin part of the formation is correlated with more than one value in the thickest part of the folded layer. For the example in Fig. 2.19, in order to calculate the mean variogram for the pairs of points (x_0, x_1) and (x_0, x_2), it is the same as doubling the sample x_0. Note that this type of transformation is possible in geological formations with a large horizontal continuity and vertical heterogeneity, which is the case in some oil and gas reservoirs.

Fig. 2.19 Structural geometric transform of a folded geological layer

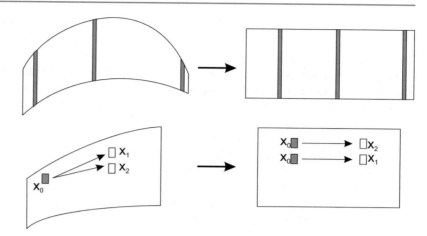

2.1.10 Co-regionalized Models of Multivariate Systems

So far we introduced the spatial continuity models for univariate systems. This section deals with a co-regionalization model for multivariate systems (Sect. 2.1.6). In a multivariate domain, the variogram models and cross-covariance, as in the univariate case, must ensure a positive variance of any linear combination between variables. There are several co-regionalization models, however the most common is the linear model. In this mode the single and cross-variograms are the result of the linear combination of basic models.

The basic models (nugget effect, first structure, second structure and so on): with L as the total number of structure, $\gamma_0(\boldsymbol{h}), \gamma_1(\boldsymbol{h}), \ldots, \gamma_L(\boldsymbol{h})$, must all belong to the simple and cross-variograms:

$$\gamma_{ii}(\boldsymbol{h}) = b_{ii}^0\gamma_0(\boldsymbol{h}) + b_{ii}^1\gamma_1(\boldsymbol{h}) + \ldots + b_{ii}^L\gamma_L(\boldsymbol{h}).$$
$$\gamma_{jj}(\boldsymbol{h}) = b_{jj}^0\gamma_0(\boldsymbol{h}) + b_{jj}^1\gamma_1(\boldsymbol{h}) + \ldots + b_{jj}^L\gamma_L(\boldsymbol{h}),$$
$$\gamma_{ij}(\boldsymbol{h}) = b_{ij}^0\gamma_0(\boldsymbol{h}) + b_{ij}^1\gamma_1(\boldsymbol{h}) + \ldots + b_{ij}^L\gamma_L(\boldsymbol{h}).$$
$$(2.46)$$

The co-regionalization model of the group of $N_v \times N_v$ simples and cross-covariance may be defined as

$$\gamma_{ij}(\boldsymbol{h}) = \sum_{l=0}^{L} b_{ij}^l\gamma_l(\boldsymbol{h}), \ \forall_{i,j},$$

or

$$C_{ij}(\boldsymbol{h}) = \sum_{l=0}^{L} b_{ij}^l C_l(\boldsymbol{h}), \quad \forall_{i,j}, \qquad (2.47)$$

where b_{ij}^l are the sill of the simple model $C_{ij}(\boldsymbol{h})$.

In order to ensure that the groups of covariances $C_{ij}(\boldsymbol{h})$ are allowed (i.e. we need to guarantee that the variance of any linear combination between variables is positive) we need the following:

(1) The functions $C_l(\boldsymbol{h})$ are positive definite;
(2) The matrices $b_{ij}^l \forall_{i,j,l=0,L}$ are positive define. This condition implies that:

$$b_{ii}^l\rangle0 \text{ e } b_{jj}^l\rangle0 \quad \text{and} \quad b_{ii}^l \cdot b_{jj}^l\rangle b_{ij}^l b_{ij}^l \ l = 0, \ldots, L.$$

These relationships suggest two basic rules:

(i) one structure that exists in the simple covariance may not exist in the cross-covariance:

$$b_{ii}^l \neq 0 \text{ does not imply that } b_{ij}^l \neq 0;$$

(ii) one structure that exists in the cross-covariance needs to necessary exist in the corresponding simple covariances:

$$b_{ij}^l \neq 0 \Rightarrow b_{ii}^l \neq 0 \text{ and } b_{jj}^l \neq 0;$$

In practice the co-regionalization models of multivariate systems by a linear model may be summarized in the following sequence of steps (Goovaerts 1997):

(1) Selection of a group of structures $\gamma_l(\boldsymbol{h}), l = 0, \ldots L$, that reproduce the behavior of the different simple variogram models $\gamma_{ii}(\boldsymbol{h})$, e $\gamma_{jj}(\boldsymbol{h})$,. Following the rules already stated in the previous paragraphs (i.e., one structure in the cross-covariance needs to exist necessarily in the corresponding simple covariances). There is no need, at this stage, to perform an analysis of the cross-variograms;

(2) Estimation of the sills b_{ij}^l for the different structures of each cross-variogram;

(3) The matrices b_{ij}^l need to be positive:

$$b_{ii}^l \rangle 0 \text{ and } b_{jj}^l \rangle 0 \quad l = 0, \ldots, L.$$

Repeat (2) and (3) until the adjustment of the group of simple and cross-variograms is satisfactory.

Note that is due to the last two steps of this sequence that in practice, make the adjustment of co-regionalization not simple.

Nevertheless, in the cases where there is the need to model multivariate systems and there is also the need to approximate the adjustments of the simple and cross-variograms, the priority should be focused on the simple variograms, and in particular, the variograms of the main variable.

2.2　Estimation Models

The reason geostatistics has grown in many research fields of Earth and Environmental Sciences relates to the efficiency and simplicity of its interpolation methods from known sparse experimental data to unknown locations within a study area.

Here we present a group of geostatistical methods for the characterization of spatial phenomena by integrating different types of experimental data with different scale support, resolution and uncertainty. The spatial inference methodologies presented for continuous and indicator variables are the basis for the stochastic sequential simulation algorithms presented in Chap. 3, and one of the main foundations of the methodologies for integrating geophysical data into reservoir modeling (Chaps. 4, 5 and 6).

2.2.1　Linear Estimation of Local Statistics

Generally speaking, it is possible to define the spatial inference, or estimation, of a variable at any given scale support (point, area or volume) for a location not sampled $Z(x_0)$, located in x_0 as a linear combination of the known value, N, for that variable $Z(x_\alpha)$ located at other different spatial positions, $x_\alpha = 1, \ldots, N$:

$$Z(\mathbf{x}_0)^* = \sum_{\alpha=1}^{n(\mathbf{u})} \lambda_\alpha Z(\mathbf{x}_\alpha). \tag{2.48}$$

The weights, λ_α, (Eq. 2.48) should summarize two extremely important characteristics in spatial inference procedures: first, they should be sensitive to the distance between the known samples $Z(\mathbf{x}_\alpha)$ and the point to be estimated $Z(\mathbf{x}_0)$; while they should also be able to disaggregate clusters of experimental data in order to avoid biasing the estimate at the unknown location, x_0, by these groups of clustered samples.

2.2.2　Probabilistic Model of the Geostatistical Linear Estimator

In the probabilistic model described in Sect. 2.1, the unknown value $Z(\mathbf{x}_0)$, and the neighboring experimental samples $Z(x_\alpha)$, $x_\alpha = 1, \ldots, N$, are interpreted as a RV located in $\mathbf{x}_0$ and $\mathbf{x}_\alpha$ and respectively.

If we assume the stationary hypothesis for the statistical moments related to the structural element bi-point (Sect. 2.1.1), the first moment of each of these RVs is individually defined as (Eq. 2.49):

$$E\{Z(x_\alpha)\} = E\{Z(x_0)\} = m. \qquad (2.49)$$

On the other hand, it is possible to assume each pair of RVs separated by the same distance vector h has the same joint distribution laws:

$$F_{ZZ'}[Z(x_1); Z(x_1 + h)] = prob\{Z(x_1) < z, Z(x_1 + h) < z'\}$$
$$= F_{ZZ'}[Z(x_2); Z(x_2 + h)]$$
$$= F_{ZZ'}[Z(x); Z(x + h)]. \qquad (2.50)$$

This means any bivariate law depends exclusively on the distance vector h—the distance between $Z(x)$ and $Z(x+h)$—and not on the location x_0. The second-order stationarity implies the variogram and covariance, $\gamma(h)$ and $C(h)$, are functions depending exclusively on the vector h.

The linear estimator described by Eq. 2.48 is interpreted as a RV located in x_0 and resulting from the linear combination of the variables $Z(x_\alpha)$, $x_\alpha = 1, \ldots, N$. Let $\varepsilon(x_0)$ be the difference between the estimated value, $Z(x_0)^*$, and the real value, $Z(x_0)$, the error associated with estimating the value of $Z(x)$ in x_0 (Eq. 2.51):

$$\varepsilon(x_0) = Z(x_0)^* - Z(x_0) = \sum_\alpha \lambda_\alpha Z(x_\alpha) - Z(x_0). \qquad (2.51)$$

The two quality criteria mentioned above may be expressed in terms of the mean and the variance of the new RV $\varepsilon(x_0)$:

(i) Unbiased condition: $E\{\varepsilon(x_0)\} = 0$

The first quality criteria of the estimate, $Z(x_\alpha)^*$, is related to its expected value:

$$E\{Z(x_\alpha)^*\} = E\{Z(x_0)\} = m. \qquad (2.52)$$

Ensuring, within the probabilistic formalism of this estimation model, there is no bias in the estimation $E\{\varepsilon(x_0)\} = 0$ results in the following:

$$E\{\varepsilon(x_0)\} = E\left\{\sum_\alpha \lambda_\alpha Z(x_\alpha)\right\} - E\left\{\sum_\alpha Z(x_0)\right\} = 0$$
$$\sum_\alpha \lambda_\alpha E\{Z(x_\alpha)\} = E\{Z(x_0)\}. \qquad (2.53)$$

Since the random function is stationary, $E\{Z(x_\alpha)\} = E\{Z(x_0)\} = m$, the equality from Eq. 2.53 is ensured if the sum of the weights is equal to 1:

$$\sum_\alpha \lambda_\alpha = 1. \qquad (2.54)$$

(ii) Variance minimization: $E\{[\varepsilon(x_0)]^2\}$

The second quality factor of this estimator is related to the variance of the error $\varepsilon(x_0)$. Two estimators may have a null mean of $\varepsilon(x_0)$, but the minimum dispersion around the mean states the difference in terms of the quality of the estimators:

$$var\{\varepsilon(x_0)\} = var\{Z(x_\alpha)^* - Z(x_0)\}$$
$$= E\left\{\left[\sum_\alpha \lambda_\alpha Z(x_\alpha) - Z(x_0)\right]^2\right\}. \qquad (2.55)$$

Decomposing the squared terms, we have:

$$= E\left\{\sum_\alpha \sum_\beta \lambda_\alpha \lambda_\beta E\{Z(x_\alpha) \cdot Z(x_\beta)\}\right\} + E\{Z(x_0)^2\}$$
$$- 2E\left\{\sum_\alpha \lambda_\alpha Z(x_\alpha) \cdot Z(x_0)\right\}$$
$$= \sum_\alpha \sum_\beta \lambda_\alpha \lambda_\beta E\{Z(x_\alpha) \cdot Z(x_\beta)\} + E\{Z(x_0)^2\}$$
$$- 2\sum_\alpha \lambda_\alpha E\{Z(x_\alpha) \cdot Z(x_0)\}. \qquad (2.56)$$

Once the covariance or variance ($C(h)$ or $\gamma(h)$), model is defined and validated for the entire study area, $Z(x_0)$, the variance of estimation of any estimator may be expressed as a function of the covariance between the samples and the unknown location at which the estimation is performed:

$$var\{\varepsilon(x_0)\} = C(0) + \sum_\alpha \sum_\beta \lambda_\alpha \lambda_\beta C(x_\alpha x_\beta)$$
$$- 2\sum_\alpha \lambda_\alpha C(x_\alpha x_\beta).$$

$$(2.57)$$

2.3 Kriging Estimate[1]

Ordinary Kriging is the most common Kriging algorithm from a family of algorithms that comprise the following non-stationary estimators: simple Kriging, universal Kriging (also known as Kriging with trend), Kriging with external drift, co-Kriging, the estimator of probability distribution functions—indicator Kriging for categorical indicator variables, and the non-linear estimates—multi-Gaussian Kriging and disjunctive Kriging (Matheron 1965).

The linear geostatistical estimator (Eq. 2.48) named by ordinary Kriging is defined as a linear combination of N neighbors of x_0—$Z(x_\alpha)$, $\alpha = 1, \ldots, N$—that verifies the two criteria related to the estimation error $\varepsilon(x_0) = Z(x_0)^* - Z(x_0)$, unbiasedness $E\{\varepsilon(x_0)\} = 0$ and minimum estimation variance (Eq. 2.58):

$$\min\{var(\varepsilon(x_0))\}. \qquad (2.58)$$

The first criterion is reached by imposing a condition on the weights (Eq. 2.54).

N partial derivatives to zero in order to λ_α ($\alpha = 1, \ldots, N$) and solving the system of N equations with N unknowns by any mathematical method. However, since the solution of the N unknowns is conditioned by Eq. 2.54, the minimization of Eq. 2.57 may be solved by resorting to Lagrange formalism, which implies adding an extra equation to Eq. 2.57 and, consequently, an extra unknown (the Lagrange parameter μ) to Eq. 2.57:

$$var\{\varepsilon(x_0)\} = C(0) + \sum_\alpha \sum_\beta \lambda_\alpha \lambda_\beta C(x_\alpha x_\beta)$$
$$- 2\sum_\alpha \lambda_\alpha C(x_\alpha x_0) + 2\mu\left[\sum_\alpha \lambda_\alpha - 1\right],$$

$$(2.59)$$

the last term is null.

The minimization of Eq. 2.59 consists of calculating the $N + 1$ partial derivatives to achieve λ_α and μ, using an equality to zero, obtaining the system of $N + 1$ equations with $N + 1$ unknowns in this way. The resulting solution is the N weights λ_α that fulfil the unbiasedness condition (Eq. 2.54) while at the same time minimizing the estimation variance:

$$\frac{\partial\left[E\left\{[Z(x_0)^* - Z(x_0)]^2\right\} + 2\mu\left[\sum_\alpha \lambda_a - 1\right]\right]}{\partial\lambda_\alpha} = 0, \alpha = 1, \ldots N$$

$$\frac{\partial\left[E\left\{[Z(x_0)^* - Z(x_0)]^2\right\} + 2\mu\left[\sum_\alpha \lambda_a - 1\right]\right]}{\partial\mu}.$$

$$(2.60)$$

The development of the N first equations results in:

$$\frac{\partial\left[C(0) + \sum_\alpha \sum_\beta \lambda_\alpha \lambda_\beta C(x_\alpha x_\beta) - 2\sum_\alpha \lambda_\alpha C(x_\alpha x_0) + 2\mu\left[\sum_\alpha \lambda_\alpha - 1\right]\right]}{\partial\lambda_\alpha} = 0, \alpha = 1, \ldots N\ 2\sum_\beta \lambda_\beta C(x_\alpha x_\beta) - 2C(x_\alpha x_0) + 2\mu = 0, \alpha = 1, \ldots N.$$

$$(2.61)$$

Minimizing the estimation variance (Eq. 2.58) is ensured by the classic method of equating the

The last partial derivative in order to μ results in the following equation:

$$\sum_\alpha \lambda_\alpha = 1. \qquad (2.62)$$

[1]The geostatistical estimator was named Kriging by Georges Matheron (1965) as a tribute to the pioneering work of Danie G. Krige (1951).

Finally, the Kriging system of $N + 1$ equations that allows calculation of the N weights λ_α is the following:

$$\begin{cases} \sum_\beta \lambda_\beta C\left(x_\alpha x_\beta\right) + \mu = C(x_\alpha x_0), \alpha = 1, \ldots N \\ \sum_\alpha \lambda_\alpha = 1. \end{cases}$$

$$(2.63)$$

The minimum estimation variance is obtained by replacing Eq. 2.63 into Eq. 2.57:

$$\sigma_E^2 = C(0) + \sum_\alpha \lambda_\alpha C(x_\alpha x_0) - \mu - 2\sum_\alpha \lambda_\alpha C(x_\alpha x_0)$$

$$\sigma_E^2 = C(0) - \sum_\alpha \lambda_\alpha C(x_\alpha x_0) - \mu.$$

$$(2.64)$$

The Kriging system may also be described in terms of the variogram $\gamma(\boldsymbol{h})$, knowing that $\gamma(\boldsymbol{h}) = C(0) - C(\boldsymbol{h})$:

$$\begin{cases} \sum_\beta \lambda_\beta \gamma\left(x_\alpha x_\beta\right) - \mu = \gamma(x_\alpha x_0), \alpha = 1, \ldots N \\ \sum_\alpha \lambda_\alpha = 1, \end{cases}$$

$$(2.65)$$

with the estimation variance defined as:

$$\sigma_E^2 = \sum_\alpha \lambda_\alpha \gamma(x_\alpha x_0) + \mu.$$

$$(2.66)$$

2.3.1 Kriging System Resolution

In practice, the system of $N + 1$ equations may be written in a matrix notation (Eq. 2.67). Considering K the covariance matrix between samples, M the second member matrix—the covariance between samples and the unknown location—and λ the weighting matrix:

$$[K] = \begin{pmatrix} C(x_1, x_1) & \cdots & C(x_1, x_N) & 1 \\ \vdots & \ddots & \vdots & \vdots \\ C(x_N, x_1) & \cdots & C(x_N, x_N) & 1 \\ 1 & 1 & & 0 \end{pmatrix},$$

$$[M] = \begin{bmatrix} C(x_1, x_0) \\ \vdots \\ C(x_N, x_N) \\ 1 \end{bmatrix} \quad [\lambda] = \begin{bmatrix} \lambda_1 \\ \vdots \\ \lambda_N \\ \mu \end{bmatrix}, \quad (2.67)$$

The Kriging system may be written as follows:

$$[K] \cdot [\lambda] = [M], \quad (2.68)$$

where the solution is achieved by inverting the matrix K:

$$[\lambda] = [K]^{-1} \cdot [M], \quad (2.69)$$

and

$$\sigma_E^2(x_0) = C(0) - [\lambda]^T \cdot [M]. \quad (2.70)$$

By defining $[Z]$ as the vector of the values $Z(x_\alpha)$, $[Z] = [z(x_1), \ldots, z(x_N)]$, the Kriging estimator $Z(x_0)^*$ is given by Eq. 2.71:

$$Z(x_0)^* = [\lambda]^T \cdot [Z] = [M]^T \cdot [K]^{-1} \cdot [Z]. \quad (2.71)$$

Note that the Kriging variance $(\sigma_E^2(x_0),$ Eq. 2.66) depends exclusively on the location of the experimental data (x_α) against the location where the estimation is being performed (x_0), and not on the values of the experimental data. In other words, the Kriging variance is not dependent on the property that is being modelled, but depends exclusively on the configuration of the experimental data against the location x_0 (Goovaerts 1997).

All the different available Kriging-based techniques share some important properties:

(1) Kriging is an exact interpolator, the values of the experimental data are honored in the interpolated model;

(2) The interpolation is constrained by a spatial continuity model, represented by a variogram model in two-point geostatistics;

(3) Kriging techniques are able to weigh differently isolated samples from clusters of samples (declustering);

(4) Models interpolated with Kriging tend to reproduce the mean value of the experimental data in areas far from the experimental data location (Deutsch and Journel 1992).

2.4 Linear Estimation of Non-stationary Phenomena: Simple Kriging

The ordinary Kriging estimator assumes the mean of the variable $Z(x)$ within the study area A is not known but constant. However, there are natural phenomena in which the values of a given property we aim to estimate are not homogeneous across the entire study area. In such cases, it can be said that a drift in the values of $Z(x)$ exists and that the stationarity hypothesis of Eq. 2.49 is not verified for the entire A.

There are some linear estimation methodologies of $Z(x)$ that take into account the way the values of $Z(x)$ drift spatially its local means $(m(x))$ vary: simple Kriging; Kriging with a trend model (universal Kriging) and Kriging with an external drift. Due to its importance in the following chapters, we will only refer to the simple Kriging estimate. For discussions of the other non-stationary Kriging estimates, the reader is referred to Journel and Huijbreghts (1978), Goovaerts (1997), and Deutsch and Journel (1992).

The simple Kriging estimate is the most general Kriging algorithm in its non-stationary version. It assumes the mean of the set of RVs from the available experimental data and the locations not sampled within the study area is known.

In practice, this algorithm is applied where the theoretical formalism of the probabilistic model imposes the knowledge of the mean of the random function (as in the multi-Gaussian Kriging of a random function with null mean) or when there is good knowledge about the trend, or drift, of the natural phenomenon. In these cases, the values of the drift (if known for the entire field) as the local mean of the RV of the sampled and non-sampled values within the study area can be assumed.

If we consider the Kriging estimator $Z(x_0)^*$ in its most general form—as a linear combination of the N data $Z(x_\alpha)$, then:

$$Z(x_0)^* = \lambda_0 \cdot 1 + \sum_{\alpha=1}^{N} \lambda_\alpha Z(x_\alpha). \qquad (2.72)$$

For those non-stationary cases in terms of the first statistical moment of the RVs are known, but not constant, the unbiasedness condition is defined as:

$$E\{Z(x_0)\} = E\{Z(x_0)^*\} = \lambda_0 + \sum_{\alpha=1}^{N} \lambda_a E\{Z(x_\alpha)\}, \qquad (2.73)$$

which implies:

$$\lambda_0 = m_{x_0} - \sum_{\alpha=1}^{N} \lambda_a m_{x_\alpha}. \qquad (2.74)$$

By plugging Eq. 2.74 into Eq. 2.72 we obtain the simple Kriging estimate:

$$Z(x_0)^* - m(x_0) = \sum_{\alpha=1}^{N} \lambda_\alpha [Z(x_\alpha) - m(x_\alpha)], \qquad (2.75)$$

in which the residual $Z(x_0) - m(x_0)$ is estimated based on the residual between samples $Z(x_\alpha) - m(x_\alpha)$.

Note that in those situations, in which uncertainty about the knowledge of the drift phenomena allows matching the drift of the set of RVs, simple Kriging is a method for estimating the residuals. The variance of the error $\varepsilon(x_0) = Z(x_0)^* - Z(x_0)$ may be written in function of the covariances (Eq. 2.57). For simple Kriging, the N weights are calculated by minimizing this variance, obtained by the N partial derivatives in order to λ_α, resulting in the following system of N equations:

$$\sum_{\beta} \lambda_\beta C(x_\alpha x_\beta) = C(x_\alpha x_0), \alpha = 1, \ldots, N, \quad (2.76)$$

resulting in the solution of N unknowns λ_α and the consequent calculus of the estimation variance associated with simple Kriging:

$$\sigma_E^2(x_0) = C(0) - \sum_\beta \lambda_\alpha C(x_\alpha x_0). \qquad (2.77)$$

2.5   Co-kriging Estimate

There are cases where, in addition to the main variable being estimated, there is a secondary variable in a different sampling grid. This secondary information may be incorporated into the estimation model as soon as a co-regionalization model between variables can be inferred. A typically example in hydrocarbon reservoir characterization is porosity, which is often modelled jointly with an existing acoustic impedance model.

Let us first consider a primary variable, $Z_1(x_i)$. $i = 1, \ldots, N_1$, known in N_1 sampling points and a secondary variable, $Z_2(x_j).j = 1, \ldots, N_2$,

$$E\{[Z_1(x_0)]_{CK}^* - Z_1(x_0)\} = 0, \qquad (2.79)$$

given the stationarity of the first statistical moment of both variables:

$$E\left\{\sum_i a_i Z_1(x_i) + \sum_j b_j Z_2(x_j) - Z_1(x_0)\right\}, \qquad (2.80)$$

$$\left[\sum_i a_i - 1\right] m_1 + \sum_j b_j m_2 = 0. \qquad (2.81)$$

The following conditions imposed over the weights ensure the unbiasedness of the estimate:

$$\sum_i a_i = 1 \ and \ \sum_j b_j = 0. \qquad (2.82)$$

The estimation variance is given by:

$$var\{\varepsilon\} = var\{[Z_1(x_0)]_{CK}^* - Z_1(x_0)\} = var\left\{\sum_{i=1}^{N_1} a_i Z_1(x_i) + \sum_{j=1}^{N_2} b_j Z_2(x_j) - Z_1(x_0)\right\}$$

$$= \sum_{i=1}^{N_1}\sum_{j=1}^{N_1} a_i a_j C_{Z_1}(x_i, x_j) + \sum_{i=1}^{N_2}\sum_{j=1}^{N_2} b_i b_j C_{Z_1}(x_i, x_j) \sum_{i=1}^{N_2}\sum_{j=1}^{N_2} b_i b_j C_{Z_2}(x_i, x_j)$$

$$+ 2\sum_{i=1}^{N_1}\sum_{j=1}^{N_2} a_i b_j C_{Z_1 Z_2}(x_i, x_j) - 2\sum_{j=1}^{N_1} a_i C_{Z_1}(x_i, x_0)$$

$$- 2\sum_{j=1}^{N_2} b_j C_{Z_1 Z_2}(x_j, x_0) + C_{Z_1}(x_0, x_0), \qquad (2.83)$$

sampled in N_2 points. The linear estimate, $Z_1(x_0)$, in an unknown location, x_0, may be described by the following linear combination of the neighbor samples of both variables $Z_1(x_i)$ and $Z_2(x_j)$:

$$[Z_1(x_0)]_{CK}^* = \sum_{i=1}^{N_1} a_i Z_1(x_i) + \sum_{j=1}^{N_2} b_j Z_2(x_j). \qquad (2.78)$$

Equation 2.78 describes the co-Kriging estimator, which like Kriging should be non-biased and with minimum variance error. The unbiasedness condition implies that the expected value for the error is null:

where $C_{Z_1}(\boldsymbol{h})$ and $C_{Z_2}(\boldsymbol{h})$ are the covariances of $Z_1(x)$ and $Z_2(x)$, respectively, and $C_{Z_1 Z_2}(\boldsymbol{h})$ the cross-covariance between $Z_1(x)$ and $Z_2(x)$. The weights a_i and b_j are calculated by minimizing the estimation variance (Eq. 2.83) with constrains from Eq. 2.82.

The Lagrange formalism may be applied to Eq. 2.83 as it was in Eq. 2.57:

$$var\{\varepsilon\} = 2\mu_1\left(\sum_{i=1}^{N_1} a_i - 1\right) + 2\mu_2\left(\sum_{j=1}^{N_2} b_j\right), \qquad (2.84)$$

where:

$$\varepsilon = [Z(x_0)]^*_{CK} - Z(x_0). \qquad (2.85)$$

The minimization of Eq. 2.83 is achieved by resorting to the $N_1 + N_2$ partial derivatives with respect to the weights a_i and b_j equal to zero:

$$\frac{\partial(var\{\varepsilon\})}{\partial a_j} = 0 \; for \; j = 1, \ldots, N_1$$

$$\frac{\partial(var\{\varepsilon\})}{\partial b_j} = 0 \; for \; j = 1, \ldots, N_2$$

$$\frac{\partial(var\{\varepsilon\})}{\partial \mu_1} = 2 \sum_{i=1}^{N_1} a_i - 1$$

$$\frac{\partial(var\{\varepsilon\})}{\partial \mu_2} = 2 \sum_{i=1}^{N_2} b_i.$$

Resulting in the $N_1 + N_2 + 2$ equations:

$$\sum_{i=1}^{N_1} a_i C_{Z_1}(x_i, x_j) + \sum_{i=1}^{N_2} b_i C_{Z_1 Z_2}(x_i, x_j) + \mu_1 = C_{Z_1}(x_0, x_j) \quad j = 1, \ldots, N_1$$

$$\sum_{i=1}^{N_1} a_i C_{Z_1 Z_2}(x_i, x_j) + \sum_{i=1}^{N_2} b_i C_{Z_2}(x_i, x_j) + \mu_2 = C_{Z_1 Z_2}(x_0, x_j) \quad j = 1, \ldots, N_2$$

$$\sum_{i=1}^{N_1} a_i = 1 \; and \; \sum_{i=1}^{N_2} b_i = 0. \qquad (2.86)$$

In sum, here are some practical notes regarding the co-Kriging estimate:

In practice, using an auxiliary variable through co-Kriging is only advantageous compared to the ordinary Kriging of the primary variable, if the primary variable is sub-sampled in comparison to the secondary variable and if both variables are correlated. For more on this topic, please see Sousa (1989), Wackernagel (1995), Bourgault and Marcotte (1991), Marcotte (1991), Myers (1982, 1984), Samper and Carrera (1990) and Goovaerts (1997).

However, there are a few points that should be noted about the co-Kriging estimate:

(1) The co-Kriging estimate is the natural extension of the Kriging estimate to incorporate a secondary variable. As in ordinary Kriging, the first member matrix must be definite positive. This condition is satisfied if the individual covariance and cross-covariance are the ones described in Sect. 2.1.8;

(2) The resolution of the system of equations for co-Kriging may have numerical instability problems if there are large differences in the variances of primary and secondary variables. In such cases, simple or cross-correlograms should be used instead of their variogram and covariance equivalents;

(3) Theoretically, the co-Kriging estimator introduces estimation errors that are smaller than those introduced by ordinary Kriging. However, this advantage must take into account the costs associated with modeling the cross-variograms, which are often just rough approximations and forced adjustments;

(4) The co-Kriging estimate may be generalized for a group of auxiliary variables:

$$[Z_1(x_0)]^*_{CK} = \sum_{\alpha_1=1}^{N_1} \lambda_{\alpha_1} Z_1(x_{\alpha_1}) + \sum_{i=2}^{N_v} \sum_{i=2}^{N_i} \lambda_{\alpha_i} Z_i(x_{\alpha_1}). \qquad (2.87)$$

The resolution of this estimate implies the knowledge of the $(N_v + 1)^2/2$ variogram models.

2.6 Co-estimation with a Secondary Variable in a Much Denser Sample Grid: Collocated Co-kriging

In some cases, the secondary variable is much more abundant compared to the available number of samples or observations of the primary variable. This is so when the secondary variable is known for the entire area A, resulting in a 2D or 3D model that is intended to be used as conditioning data to estimate the primary variable. Note that the term model here refers to knowledge of the variable for the entire study area. When this happens, the co-Kriging system (Eq. 2.86) becomes unstable due to difference in the sampling density between the primary and secondary variables.

Moreover, the value of the secondary variable at the location at which the estimation is performed tends to minimize the effect from the primary variable samples located at great distances. In these cases, one possible solution is to retain the values of the secondary variable at the location being studied. This Kriging version is known as collocated co-Kriging (Xu et al. 1992; Almeida and Journel 1994).

By considering $Z_1(x)$ to be the primary variable the values of which are known at N_1 locations, and $Z_2(x)$ the secondary variable known for the entire study area, the collocated co-Kriging estimate $Z_1(x_0)$ is defined by:

$$[Z_1(x_0)]_{CK}^* = \sum_{\alpha_1=1}^{N_1} a_i Z_1(x_i) + b_0 Z_2(x_0). \quad (2.88)$$

Resulting in a system of equation of $N_1 + 2$ equations:

When the variances of the primary and secondary variables are very different, the system of equations (Eq. 2.89) should be expressed in terms of correlograms.

2.7 Estimation of Local Probability Distribution Functions

The Kriging estimate presented above is an optimal solution for inferring mean global or local characteristics of a quantitative property. The resulting models are interpolated models of mean values that are traditionally suitable for characterizing variables homogenously spatially distributed, i.e. variables with low variability in which the mean value is enough to represent it within a study area.

For heterogeneous variables, such as the internal petro-elastic properties of a hydrocarbon reservoir, since Kriging results in smooth models, the Kriging estimate is not enough to characterize their spatial distribution.

For such complex variables there are geostatistical models that aim to locally estimate the probability distribution function of a given property. These probability distribution functions are the basis for the sequential stochastic simulation methodologies introduced in the Chap. 3.

These can be used in the context of this book to map extreme values or to assess local uncertainty (Goovaerts 1997). But the indicator formalism as a method for estimating local probability distribution functions is based in the work developed by Switzer (1977). However, this method is difficult to implement and was replaced by alternative stochastic simulation processes for continuous variables, such as the

$$\sum_{i=1}^{N_1} a_i C_{Z_1}(x_i, x_j) + b_0 C_{Z_1 Z_2}(x_i, x_0) + \mu_1 = C_{Z_1}(x_0, x_j) \quad j = 1, \ldots, N_1$$

$$\sum_{i=1}^{N_1} a_i C_{Z_1 Z_2}(x_i, x_j) + b_0 C_{Z_2}(0) + \mu_2 = C_{Z_1 Z_2}(0) \quad j = 1, \ldots, N_1$$

$$\sum_{i=1}^{N_1} a_i + b_0 = 1.. \quad (2.89)$$

Gaussian formalism. Readers interested in the indicator formalism should read Goovaerts (1997) and Deutsch and Journel (1992).

Multi-Gaussian formalism

This approach for estimating the local probability functions of a given variable consists in using a single model known for the distribution function of the group of RVs. It assumes that a group of RVs $\{Y(x), x \in A\}$ follows a joint multi-Gaussian function. This is an easy way to estimate the local probability distribution function when compared to the indicator formalism, but this strong assumption may lead to some consistency problems with the available experimental data.

The probability distribution function at any location x_0 is perfectly described by the conditional expected value and variance:

$$E\{Y(x_0)|Y(x_1)\ldots Y(x_N)\},$$
$$var\{Y(x_0)|Y(x_1)\ldots Y(x_N)\}. \qquad (2.90)$$

Resulting in the Gaussian probability function at x_0 as:

$$G(x_0; y) = G\left[\frac{y - E\{Y(x_0)|Y(x_\alpha), \alpha = 1, \ldots, N\}}{\sqrt{var\{Y(x_0)|Y(x_\alpha), \alpha = 1, \ldots, N\}}}\right].$$
$$(2.91)$$

Under the multi-Gaussian assumption both first statistical moments (Eq. 2.90) are equal to the linear simple Kriging estimate (Eq. 2.75) and the corresponding Kriging variance (Eq. 2.77; Journel and Huijbreghts 1978):

$$E\{Y(x_0)|Y(x_\alpha), \alpha = 1, \ldots, N\} = [Y(x_0)^*]$$
$$= m(x_0) + \sum_{\alpha=1}^{n} \lambda_\alpha [Y(x_\alpha) - m(x_\alpha)] = \sum_{\alpha=1}^{N} \lambda_\alpha Y(x_\alpha),$$
$$(2.92)$$

given that the means are known and constant:

$$m(x_0) = m(x_\alpha) = 0,$$
$$var\{Y(x_0)|Y(x_\alpha), \alpha = 1, \ldots, N\} = \sigma_E^2(x_0).$$
$$(2.93)$$

where the weights λ_α are computed by the simple Kriging system (Eq. 2.76).

The probability distribution function in x_0 is defined by the two parameters estimated by simple Kriging—mean and variance:

$$G(x_0; y) = G\left[\frac{y - [Y(x_0)^*]}{\sigma_E^2(x_0)}\right]. \qquad (2.94)$$

2.7.1 Gaussian Transform of the Experimental Data

One of the greatest advantages of this approach concerns the simplicity of its implementation: the probability distribution function is defined for every location x_0 with the simple Kriging estimate of $Y(x_0)$. However, we need a Gaussian transformation of the experimental data $Z(x_\alpha)$, $\alpha = 1, \ldots, N$ to ensure the Gaussian marginal distribution at least:

$$Y(x_\alpha) = \Phi(Z(x_\alpha)), \alpha = 1, \ldots, N, \qquad (2.95)$$

where $Y(x_\alpha)$ follows a Gaussian function with zero mean and variance one.

The Gaussian transform (Φ) may be calculated using a polynomial approximation—Hermite's polynomial (Matheron 1974; Muge 1982)—or by a simple graphical transform, which due to its simplicity is more suitable for this operation.

Given two distribution function of variables $Z(x)$ and $Y(x)$:

$$F(x) = prob\{Z(x) < z\}.$$
$$G(y) = prob\{Y(x) < y\},$$

the value z corresponding to the Gaussian value y satisfies $F(z) = G(y)$.
Generalizing:

$$Y(x_\alpha) = \Phi(Z(x_\alpha)) = G^{-1}(F(Z(x_\alpha)))$$
$$\alpha = 1, \ldots, N, \qquad (2.96)$$

with the Gaussian transformation of the experimental data $Z(x)$, with a probability distribution function $F(z)$ into Gaussian $Y(x)$, and assuming that these follow a joint multi-Gaussian probability function, all the formalisms previously described may be applied following this sequential approach:

(1) The experimental data is transformed into Gaussian (Eq. 2.96);
(2) After calculating the variograms of the transformed values $Y(x_\alpha)$ for each single point x_0, a local probability distribution function is calculated:

$$G(x_0; y) = prob\{Y(x_0) < y\}; \qquad (2.97)$$

(3) The values of the probability distribution $F(z')$ for any threshold value z' are obtained by the inverse transform ϕ: first the value of y' corresponding to z' is calculated:

$$y' = \phi(z') = G^{-1}[F(z')]. \qquad (2.98)$$

Then, we may calculate $F(x_0, z')$ from $G(x_0; y')$ estimated by Eq. 2.94:

$$F(x_0, z') = G(x_0; y'). \qquad (2.99)$$

If the threshold z' is not coincident with the values of experimental data $Z(x_\alpha)$, and because $F(x, z)$ is monotonous crescent, the inverse transformation may be calculated by a linear or power interpolation (Goovaerts 1997).

2.8   Estimation of Categorical Variables

When the petrophysical properties of a given hydrocarbon reservoir have some degree of homogenization, one can frequently classify the group of petrophysical properties through the concept of lithofacies. Lithofacies may be defined as geological bodies that share an identical behavior in terms of petrophysical and/or elastic response; this concept does not strictly refer to distinct types of lithologies (or sedimentary facies). In reservoir characterization, these lithofacies are frequently modelled using categorical variables, which are further modelled in terms of their internal properties and to continuous variables, such as porosity and permeability.

Within the conceptual geostatistical framework for categorical variables, the unit element consists in the probability of a point located within the study area that belongs to a group of complementary and disjunctive bodies. The shapes of the different bodies (or lithofacies) result from the classification of these elements with the greater probability of belonging to each body (or lithofacies). Let's assume a group K of disjunctive lithofacies, $X_k = 1, \ldots, K$. For each point located in x within the study area A we may define a binary vector $I_k(x)$ as follows (Fig. 2.20):

$$I_k(x) = \begin{cases} 1, & \text{if } x \in X_k \\ 0, & \text{if } x \in X_j \text{ and } j \neq k. \end{cases} \qquad (2.100)$$

Fig. 2.20 Schematic representation of a group of three lithofacies in a 2D model

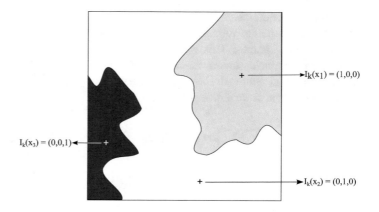

From the multi-phase group, we may define the individual statistics for each single phase:

$$m_k = E\{I_k(x)\},$$
$$\sigma_k^2 = var\{I_k(x)\}. \tag{2.101}$$

For the group of different lithofacies we may define a measurement of average continuity of global structure—$C(\mathbf{h})$ (Eq. 2.102)—with the probability of two points x and $x + \mathbf{h}$ with distance $\mathbf{h}$, belonging to the same lithofacies X_k for all $k = 1, ..., K$ (Soares 1992):

$$C(\mathbf{h}) = E\left\{\sum_{k=1}^{K} I_k(x) \cdot I_k(x+\mathbf{h})\right\}. \tag{2.102}$$

The multi-phase covariance (Eq. 2.102) may be decomposed by the sum of the single-phase covariance:

$$C(\mathbf{h}) = \sum_{k=1}^{K} E\{I_k(x) \cdot I_k(x+\mathbf{h})\} = \sum_{k=1}^{K} C_k(\mathbf{h}), \tag{2.103}$$

which may be written in the form of a variogram.

From the structural point of view, the covariance $C(\mathbf{h})$ and the variogram $\gamma(\mathbf{h})$ quantify the average morphological variability of the set of all lithofacies:

$$\gamma(\mathbf{h}) = \frac{1}{2}E\left\{\sum_{k=1}^{K}[I_k(x) - I_{ki}(x+\mathbf{h})]^2\right\}. \tag{2.104}$$

Note that for most cases, when dealing with multi-phase structures there are not enough samples to estimate reliable individual variograms or covariances. This gets worse as the number of lithofacies increases. In this way the multi-phase variogram or covariance may be directly estimated from Eq. 2.105:

$$\gamma(h) = \frac{1}{KN(h)} \sum_{k=1}^{K} \sum_{\alpha=1}^{N(h)} [I_k(x_a) - I_{ki}(x_{\alpha+\mathbf{h}})]^2,$$

$$C(h) = \sum_{i=1}^{K} \sigma_k^2 - \gamma(\mathbf{h}). \tag{2.105}$$

A multi-phase group may be composed of subgroups with distinct spatial behavior. This can be seen as having a single variable with a non-stationary behavior for the entire study area. By adopting an average global spatial continuity model there will be regional areas that are not modelled correctly. In such cases, each subgroup should be modelled independently, avoiding as much as possible, the integration within the same model subgroups with very distinct spatial behaviors.

With these structural tools for categorical variables, we are now ready to introduce a morphological estimation methodology for multi-phase structures (Soares 1992). With the geostatistical estimation of multi-phase structures, the aim is to calculate, for each point x_0 within the study area A, the joint probability of x_0 belonging to lithofacies X_k for all $k = 1, ..., K$ based on the $I_i(x_\alpha)$ from the experimental samples $x_\alpha = 1, ..., N$:

$$prob\{x_0 \in X_1\} = [I_1(x_0)]^* = \sum_\alpha \lambda_{\alpha,1} I_1(x_\alpha),$$

$$prob\{x_0 \in X_2\} = [I_2(x_0)]^* = \sum_\alpha \lambda_{\alpha,2} I_2(x_\alpha),$$

$$prob\{x_0 \in X_K\} = [I_K(x_0)]^* = \sum_\alpha \lambda_{\alpha,K} I_K(x_\alpha). \tag{2.106}$$

If when constructing these estimators, we use the same covariance model—multi-phase covariance—then the weights λ_α are the same for all phases:

$$\lambda_{\alpha,1} = \lambda_{\alpha,2} = \ldots = \lambda_{\alpha,K} = \lambda_\alpha. \tag{2.107}$$

The estimator $[I_K(x_0)]^* = \sum \lambda_\alpha I_k(x_\alpha)$, $k = 1, \ldots, K$, is calculated in such a way that it is not biased and the estimation variance is minimized:

Unbiasedness condition

$$E\{[I_k(x_0)]^*\} = E\{I_k(x_0)\},$$

$$\sum_\alpha \lambda_\alpha E\{I_k(x_\alpha)\} = E\{I_k(x_0)\}, \tag{2.108}$$

implies that

$$\sum_\alpha \lambda_\alpha = 1. \qquad (2.109)$$

Consequently, from Eq. 2.106 the result is that the sum of the estimated probabilities of belonging to each one of the phases is 1:

$$\sum_{k=1}^{K} [I_k(x_0)]^* = \sum_\alpha \lambda_\alpha = 1. \qquad (2.110)$$

In addition to the unbiasedness condition, Eq. 2.108 calculated the weights of the minimization in the sum of the estimation variances that is also imposed on each lithofacies individually:

$$min\left\{ E[[I_1(x_0)]^* - I_1(x_0)]^2 + E[[I_2(x_0)]^* - I_2(x_0)]^2 + \ldots + E[[I_K(x_0)]^* - I_K(x_0)]^2 \right\}$$
$$= min\left\{ E\left[\sum_k E[[I_k(x_0)]^* - I_k(x_0)]^2 \right] \right\} \qquad (2.111)$$

which we also may express in terms of multi-phase covariance $C(h)$:

$$\sum_k E\left\{ [[I_k(x_0)]^* - I_k(x_0)]^2 \right\}$$
$$= C(0) + \sum_\alpha \sum_\beta \lambda_\alpha \lambda_\beta C(x_\alpha x_\beta) + \sum_\alpha \lambda_\alpha C(x_\alpha x_0). \qquad (2.112)$$

By minimizing this equation under the constraining from Eq. 2.109, we obtain the classical Kriging system with multi-phase covariance:

$$\begin{cases} \sum_\beta \lambda_\beta C(x_\alpha x_\beta) + \mu = C(x_\beta x_0) & \forall \alpha = 1, \ldots, N \\ \sum_\alpha \lambda_\alpha = 1. \end{cases} \qquad (2.113)$$

The minimization of the sum of the variances does not directly imply the minimization of the variances for each class individually. This means that the best estimate of the multi-phase group may not be the best estimate of each class/lithofacies individually: this is only possible if each class is estimated individually with independent covariance for each class.

When we can calculate individual variograms for each phase, they can be different, and as the weights are phase-dependent, the sum of probabilities estimated at a given point may not be 1. However, there are methods to overcome this limitation (Suro-Perez and Journel 1990).

Considering:

$$S_i = \sum_\alpha [I_{ki}(x)]^* \neq 1, \qquad (2.114)$$

then the estimated probability for each phase is reconverted by factor

$$[I_K(x_0)]^{**} = \frac{[I_K(x_0)]^*}{S_i} \quad being \quad \sum_k [I_k(x_0)]^{**} = 1.$$

Thus, both the multi-phase estimation and the individual estimation of each phase are valid methods for reaching the same goal—the spatial characterization of a multi-phase structure—but in different situations:

- The individual estimation of each phase has the advantage of taking into consideration the structural differences quantified by individual variogram or covariance models. However, in practice this is not often used since while the number of phases increase, the estimation of the individual variograms becomes more difficult as the number of samples per phase decreases.
- Using multi-phase variograms does not mean we need to use a single variogram model. Heterogeneous multi-phase groups may and should be modelled by different multi-phase variograms within the same estimation procedure.

Simulation Models of Physical Phenomena in Earth Sciences

3

3.1 Stochastic Simulation Models

The Kriging models presented in Sect. 2.2 above can be considered smooth representations of subsurface geology. The resulting models represent the expected mean values of the property of interest and the extreme low and high values are usually underestimated (Fig. 3.1) (Deutsch and Journel 1992; Soares 2006). However, for reservoir characterization, and in particular for reserve calculations and fluid flow simulation, it is important to make decisions based on reliable and highly detailed subsurface Earth models. These models should be geologically realistic, incorporating small-scale variability, and able to reproduce the extreme values—both low and high extremes—of the experimental data. Unless the variable that is being modelled is very homogeneous (e.g. top, bottom and thickness of layers for some geological environments), a smooth representation of the subsurface geology is clearly insufficient for assessing the uncertainty and effects of extreme scenarios (e.g. very high and low porosity values in fluid simulation and history matching problems). The answer to these challenges, in particular in hydrocarbon reservoir characterization, has been given by geostatistics through stochastic sequential simulation models (Goovaerts 1997; Deutsch and Journel 1992).

In fact, the estimation methods presented here can infer local statistical parameters (mean and variance) of local cumulative distribution functions (cdfs), which are the basis for the stochastic simulation models presented in this section, and for the assessment of the spatial uncertainty and extreme values by generating multiple spatial correlated realizations of the study's main attributes.

The geostatistical simulation models interpolate the property of interest, reproducing marginal and joint distributions (joint simulation) as retrieved from the experimental data. They also ensure the reproduction of the spatial continuity pattern as imposed by a variogram model and allow the assessment of the spatial uncertainty related with a given property.

For example, Fig. 3.1 shows a 2D model that was created Kriging experimental data from P-wave velocity, representing the average behavior of that parameter. The other three figures are simulated models of the same property resorting to stochastic sequential simulation. Each individual model can be thought of as a possible and equiprobable outcome RF (Sect. 2.1.3). A set of realizations comprises simulated Earth models that differ from each other due to the intrinsic properties of stochastic simulation algorithms, but are considered equiprobable under the same assumptions about the prior probability distributions and the spatial continuity model. All the models belonging to an ensemble of realizations share the same properties: the reproduction of the values of the experimental data at its location; the reproduction of the prior probability distribution as estimated

© Springer International Publishing AG 2017
L. Azevedo and A. Soares, *Geostatistical Methods for Reservoir Geophysics*,
Advances in Oil and Gas Exploration & Production, DOI 10.1007/978-3-319-53201-1_3

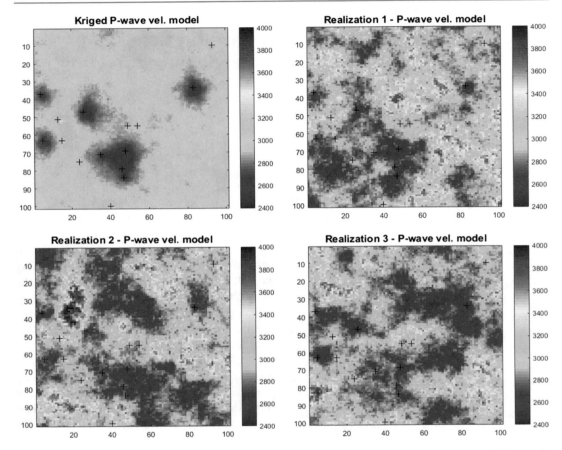

Fig. 3.1 Comparison between a Krigged model (*upper left corner*) with three different realizations for P-wave velocity. The experimental data used in the modeling process is the same for all the four models and its location is represented by the black crosses. The variability in the simulated models is much higher when compared to the interpolated model with Kriging

from the experimental data; and the reproduction of a spatial continuity model as revealed by a variogram model, for example.

In the context of this book, the objectives of the simulation models are two-fold:

– Simulated models are the privileged geostatistical tool for characterizing the spatial dispersion and spatial uncertainty for the petrophysical properties, facies, etc. of hydrocarbon reservoirs;
– The geostatistical seismic inversion methodologies (presented in Chaps. 4, 5 and 6) are iterative optimization methods based on a crucial step of generation/perturbation of

petrophysical and petro-elastic models of the reservoir—porosity, velocities, acoustic and/or elastic impedances. In the presented methodologies, this step is performed with stochastic sequential simulations and joint simulations of these properties.

Since the pioneering work of Journel (1974), the Turning Bands simulation method with independent conditioning, the use of stochastic simulations for modeling and assessing the spatial uncertainty of geological phenomena has been the object of different geostatistical simulation methods: Borgman et al. (1984), Davis (1987), Srivastava (1992), Froidevaux (1993),

Omre et al. (1993), Armstrong and Dowd (1994), Shrivastava (1995), Lantuejoul (2002), Dimitrakopulos and Luo (2004), Richmond and Dimitrakopulos (2005) and Boucher and Dimitrakopulos (2007). This book focuses on a family of simulation algorithms, the stochastic sequential simulation methods (Alabert 1987; Journel and Alabert 1988; Gómez-Hernández and Srivastava 1990) and, in particular, the direct sequential simulation (Soares 2001), given its suitability for the purpose of the geostatistical stochastic inversion.

3.2 Sequential Simulation Models

The basic principle behind sequential simulation is extremely simple and is based on the application of Bayes' rule in successive sequential steps (Law and Kelton 1991; Ripley 1987; Journel and Alabert 1988). Let us assume we wish to generate a set of values $z_1, z_2, ..., z_N$ from a multivariate distribution $F(Z_1, Z_2, Z_3, ..., Z_N)$. This can be expressed by the Bayes' relation:

$$F(Z_1, Z_2, Z_3, ..., Z_N) = F(Z_1)F(Z_2|Z_1)F(Z_3|Z_1, Z_2)$$
$$...F(Z_N|Z_1, Z_2, Z_3, ..., Z_{N-1}). \tag{3.1}$$

The first value z_1 is sampled from $F(Z_1)$ by, for example, Monte Carlo inverse transform algorithm (Law and Kelton 1991); the value z_2 is sampled from the conditional distribution $F(Z_2|Z_1 = z_1)$; the last value z_N is sampled from the conditional distribution $F(Z_N|Z_1, Z_2, Z_3, ..., Z_{N-1})$. Hence, the set of values $z_1, z_2, ..., z_N$ follows the joint distribution $F(Z_1, Z_2, Z_3, ..., Z_N)$.

The N dependent variables $Z_1, Z_2, Z_3, ..., Z_N$ can represent the same property spatially located at the N nodes of a regular grid covering the study area. With n conditioning values corresponding to the experimental well samples data $Z_\alpha, \alpha = 1, n$, the joint distribution can be written as $F(N) = (Z(x_1), Z(x_2), Z(x_3), ..., Z(x_N)|(n))$.

The crucial point of the sequential simulation methodologies is precisely the knowledge of N conditional cumulative distribution functions:

$$prob\{Z(x_1) < z|(n)\}$$
$$prob\{Z(x_2) < z|(n+1)\}$$
$$\vdots \tag{3.2}$$
$$prob\{Z(x_N) < z|(n+N-1)\}.$$

Journel and Alabert (1988) proposed the use of geostatistics for a local estimation of those functions. Two geostatistical algorithms can be used to estimate these local conditional distributions based on the indicator and multi-Gaussian Kriging (Chap. 2): sequential Gaussian simulation (Sect. 3.3) and sequential indicator simulation to simulate continuous and categorical variables (Sect. 3.5), respectively.

The third simulation algorithm presented in this book, direct sequential simulation (Sect. 3.4), does not require the simulated variable to be transformed. Local conditional mean and variance of Z(x), estimated by simple Kriging, are used for the re-sample of the global distribution function of Z(x). As this method works in the space of the original variables, this property creates a high potential for accommodating complex relationships between covariates. In this chapter, joint simulations with reproduction of bivariate distributions and point distributions are also presented (Sect. 3.4.3).

3.3 Sequential Gaussian Simulation

In sequential Gaussian simulation (SGS) the entire procedure is developed with a multi-Gaussian distributed variable. For this reason, the first step of this sequential methodology is the transformation of the original data from the available experimental data into a marginal Gaussian distributed variable:

$$Y(x) = \varphi[Z(x)]. \tag{3.3}$$

This marginal Gaussian distributed $Y(x)$ is assumed to be multi-Gaussian (see the multi-Gaussian approach in Chap. 2).

Assuming $Y(x)$ is multi-Gaussian, the conditional probability density function at any location x_0 is fully characterized by the conditional mean and variance, which can be calculated by a linear estimation, such as a simple Kriging estimate (Eq. 2.75) and variance (Eq. 2.77) (Matheron 1978). The sequential Gaussian simulation methodology can then be summarized in the following sequence of steps:

(1) Transform the original experimental data into Gaussian by applying Gaussian transform (Eq. 3.3);

(2) Generate a random path over the entire simulation grid of nodes $x_u, u = 1, \ldots, N$, where N is the total number of nodes that compose the regular grid;

(3) Estimate the local mean, $z(x_0)^*$, and variance, $\sigma_{SK}^2(x_0)$, with simple Kriging estimates conditioned to the original experimental data and previously simulated data;

(4) Draw a value y^s from the Gaussian cumulative distribution $G(y(x_0)^*, \sigma_{SK}^2(x_0))$;

(5) Add the simulated value to the conditioning data;

(6) Loop until all the N nodes of the simulated grid have been simulated;

(7) Inverse the Gaussian transform to convert the simulated values to the original data domain $z^s(x) = \varphi^{-1}(y_x^s)$.

This process of sequential simulation theoretically ensures the reproduction of the variogram model of $Y(x)$ (Journel 1989). However, the final back transform can be a critical step: by imposing the original histogram $Z(x)$ when this is considerably different compared with the bell-shaped and simetric Gaussian distribution, it can jeopardize the variogram model to be reproduced.

The sequential Gaussian simulation, as the other sequential simulation algorithms presented in this book, contains a practical approximation: in each step of the simulation process the simulated value is included in the set of conditioning data and as the conditioning values increase a linear combination of n conditioning values can be difficult to calculate accurately (Gómez-Hernández and Journel 1993). One approximation consists of selecting a limited set of conditioning data $n_1 \subset n$, such that $n_1 \ll n$ and $E\{Y(x_0)|(n_1)\} \approx E\{Y(x_0)|(n)\}$.

The selection of the subsets of conditioning data around the location x_0 must be careful. If the selection of the conditioning data promotes samples near the location (x_0) to be simulated it results in the preferential reproduction of small-scale structures in the simulated model. On the other hand, if the searching radius for conditioning data is too large it may cause a false homogeneity in the resulting simulated models.

3.4 Direct Sequential Simulation from Experimental Distributions

This section presents a detailed description of a set of stochastic simulation algorithms that are directly sampled from experimental distribution functions, marginal distributions or bi-distributions. Direct sequential simulation is a stochastic sequential simulation methodology that does not require any nonlinear transformation of the experimental data [e.g. Gaussian transform (Sect. 3.3)]. For this reason, direct sequential simulation is a suitable method for reproducing complex relationships between variables, such as the acoustic and elastic properties and rock physics properties. In addition, it is a preferable stochastic simulation tool for dealing with local distributions and local variogram models for different spatial areas within a given field (see the case studies in Chaps. 4 and 5). These important achievements make this set of simulation methods appropriate for the seismic inversion algorithms discussed in Chaps. 4, 5 and 6.

An executable of the Direct Sequential Simulation algorithm with examples of application for all the methodologies presented in the next sections is available for download at: https://sites.google.com/view/directsequentialsimulation.

3.4.1 Direct Sequential Simulation

Sequential Gaussian simulation presents some limitations when seeking to reproduce more complex structures, such as those with highly skewed or multi-modal distributions. Retrieving the original probability distribution function after the inverse Gaussian transform can lead to a more unstructured variogram. Identical problems with variogram reproduction usually are faced when one wishes to reproduce the joint statistics of a multivariate set of data.

The idea of the direct use of original variables was launched based on the following idea (Journel 1994; Journel and Xu 1994): following the traditional sequential approach, if the local conditional distributions are centered in a simple Kriging estimate and have a conditional variance equal to the simple Kriging variance, then the variogram model reproduction is assured. The problem is that the reproduction of histograms as retrieved from original variables, is not assured. In other words, the local conditional distribution functions cannot be fully characterized only by the conditional mean and variance of the original variables.

Soares (2001) proposes a direct sequential simulation (DSS) algorithm, based on the resample of the global distribution conditioned by the local conditional mean and variance. The concept of resampling the global experimental distributions was extended to bi-distributions (co-simulation of a multivariate set, Sect. 3.4.3) and distributions of uncertainty data (simulation of point distribution, Sect. 3.4.4), methods presented in this chapter. These are clearly the advantages of using DSS algorithms as the core of the inverse methods presented in the next chapter, since they ensure the reproduction of the marginal and joint distribution of the reservoirs' properties being modelled.

The DSS algorithm follows the sequential simulation approach as introduced above. When comparing with other sequential simulation algorithms, such as the SGS, instead of using the simple Kriging estimate (Eq. 2.75) and the simple Kriging estimation variance (Eq. 2.77) to define the local conditional distributions from where the simulated value is drawn, DSS uses the estimated local mean and variance to sample directly from the global conditional distribution function as estimated from the experimental data (Fig. 3.2; Soares 2001). The simulated values are drawn from an auxiliary probability distribution function ($F'_z(z)$) that is built from the global cumulative distribution function $F_z(z)$. $F'_z(z)$ is defined by selecting an interval over $F_z(z)$ centred on the simple Kriging estimate $(z(x_0)^*)$ value with an interval range proportional to the Kriging variance, σ^2_{sk} (Soares 2001):

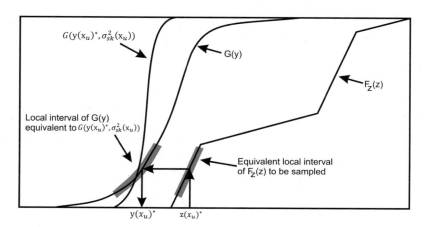

Fig. 3.2 Sampling of global probability distribution $F_z(z)$ by intervals defined by the local mean and variance of z(x_0): The value $\mathbf{y}(x_0)^*$ corresponds to the local estimate $z(x_0)^*$. The simulated value $z(x_0)^*$ is drawn from the interval of $F_z(z)$ defined by $G(\mathbf{y}(x_0)^*, \sigma^2_{SK}(x_0))$ (adapted from Soares 2001)

$$\frac{1}{n}\sum_{i=1}^{n} z(x_i) = [z(x_0)]^*, \qquad (3.4)$$

$$\frac{1}{n}\sum_{i=1}^{n}[z(x_i) - [z(x_0)]^*]^2 = \sigma_{sk}^2(x_0). \qquad (3.5)$$

One way of constructing $F'_z(z)$ is by defining a local Gaussian cumulative distribution function $G(y(x_0)^*, \sigma_{SK}^2(x_0))$, created by the Gaussian transform (Eq. 3.3) of the interval of $F_z(z)$ centered in $z(x_0)^*$ with an amplitude proportional to σ_{sk}^2:

$$y(x) = \varphi(z(x)), \quad \text{with } G(y(x)) = F_z(z(x)). \qquad (3.6)$$

The procedure to simulate a given $(z^s(x_0))$ is illustrated in Fig. 3.2. First, the interval of the original global probability distribution defined by the local simple Kriging estimate $z(x_0)^*$ and variance $(\sigma_{SK}^2(x_0))$ is transformed into the Gaussian domain $(G(y(x_0)^*, \sigma_{SK}^2(x_{x_0})))$, then a value y^s is drawn from the interval defined by $G(y(x_0)^*, \sigma_{SK}^2(x_0))$ by Monte Carlo simulation. Finally, the simulated value $(z^s(x_0))$ is obtained by the inverse transform: $z^s(x_0) = \varphi^{-1}(y^s)$.

It is important to highlight that the Gaussian transform is exclusively used to resample the intervals of $F_z(z)$ and that there is no need to assume Gaussian hypothesis for the original variable as in SGS (Soares 2001).

The DSS simulation algorithm can be summarized in the following sequence of steps (Soares 2001):

(1) Generate a random seed to define a random path over the entire simulation grid $x_u, u = 1, \ldots, N$, where N is the total number of nodes that compose the simulation grid;
(2) Estimate the local mean, $z(x_0)^*$, and variance, $\sigma_{SK}^2(x_0)$, with simple Kriging estimate conditioned to the original experimental data and previously simulated data, within a neighborhood around u;

(3) Define the interval of the global $F_z(z)$ to be sampled (Fig. 3.2);
(4) Draw a value $z^s(x_0)$ from the cumulative distribution function of $F_z(z)$;
    Generate a value u from the uniform distribution between [0, 1];
    Generate a value y^s from $G(y(x_0)^*, \sigma_{SK}^2(x_u))$;
    Return the simulated value $z^s(x_0) = \varphi^{-1}(y^s)$;
(5) Add the simulated value into the conditioning data;
(6) Loop until all the N nodes of the simulated grid have been simulated.

The resulting simulated models are able to reproduce the prior probability distribution as estimated from the experimental data. They honor the data values at its own locations and are able to reproduce the spatial continuity pattern imposed by a covariance model, i.e. the variogram model (Fig. 3.3).

3.4.2 Direct Sequential Co-simulation

In many situations we want to generate spatial realizations of two or more properties (e.g. porosity, acoustic impedance, permeability) by reproducing the underlying correlation that may eventually exist between them. This can be achieved by joint simulations or co-simulation models (Soares 2001; Horta and Soares 2010). This book focuses on the class of sequential co-simulations, and in particular the direct sequential co-simulation.

The Bayes' principle (Eq. 3.1) can be extended for the co-simulation procedure: first a variable is simulated; then a second variable is co-simulated conditioned to the first. For example, assuming a case in which we have two dependent variables: a secondary variable $Z_1(x)$, which is first simulated by any stochastic sequential simulation procedure (e.g. SGS, DSS); and a primary variable $Z_2(x)$, the new property that we want to co-simulate, conditioned to the previously simulated model of $Z_1(x)$.

In the direct sequential co-simulation, in order to sample from the global prior

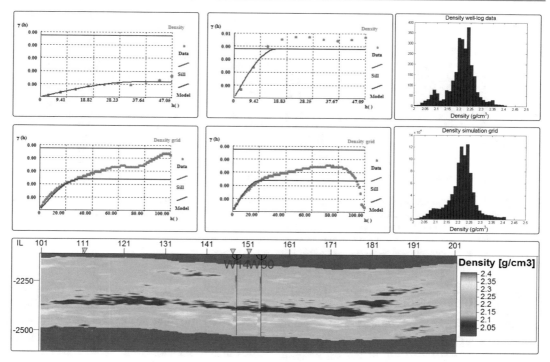

Fig. 3.3 *Top* Omnidirectional, and vertical variogram models and probability distribution function estimated from the well-log data. *Middle* Omnidirectional, and vertical variogram models and probability distribution function estimated from a realization simulated with DSS using the well-log from the top figures as conditioning data. The variogram models and the probability distribution functions, as estimated from the experimental data, are reproduced in the simulated models. *Bottom* Vertical section extracted from the simulated model honoring the well-log data (W14 and W30) at its location

probability distribution of $F_{Z_2}(z)$, we estimate the local mean and variance of $Z_2(x)$, conditioned not only to the $Z_2(x)$ sample data but also to the collocated datum at x_0, of $Z_1(x)$. This can be achieved by using a collocated simple co-Kriging estimate (Almeida and Journel 1994; Soares 2001):

$$[Z_2(x_0)^*]_{CSK} = \sum_{\alpha=1}^{N} \lambda_\alpha [z_2(x_\alpha) - m_2] \\ + \lambda_\beta [z_1(x_0)^* - m_1] + m_2, \quad (3.7)$$

with m_1 and m_2 the mean of $z_1(x)$ and $z_2(x)$, respectively,

$$\sigma_{CSK}^2(x_0) = \mathrm{Var}\{Z_2(x_0)^* - Z_2(x_0)\}. \quad (3.8)$$

The co-simulation process can be summarized in the following sequence of steps:

(1) Simulate for the grid the first variable $Z_1(x)$ with DSS;

(2) Generate a random path over the entire simulation grid $x_0, u = 1, \ldots, N$, where N is the total number of nodes that compose the simulation grid;

(3) Estimate the local mean and variance at x_0 with collocated simple co-Kriging estimate ($[Z_2(x_0)^*]_{CSK}$) and the corresponding co-Kriging variance ($\sigma_{CSK}^2(x_0)$) conditioned to the neighborhood data ($z_2(x_\alpha)$), composed by the experimental and the previously simulated data, and the colocated datum ($z_1(x_0)$);

(4) Define the interval of $F_{z2}(z)$ to be sampled, as previously explained (Fig. 3.2);

(5) Draw a value $z_2(x_0)^s$ from the cumulative distribution function of $F_{z2}(z)$:
Generate a value u from the uniform distribution between [0, 1];

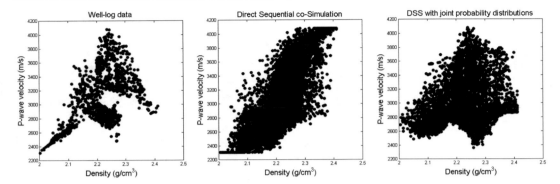

Fig. 3.4 *On the left* Joint distributions estimated from well-log data, *middle* joint distributions estimated from the resulting bi-distributions between co-simulated models using direct sequential co-simulation and *on the right* from direct sequential co-simulation with joint probability distributions. The use of DSS with joint probability distributions allows the reproduction of the bi-distributions estimated from the well-log data in the simulated models

Generate a value y^s from $G(y(x_0)^*, \sigma^2_{CSK}(x_0))$;
Return the simulated value $z_2^s(x_0) = \varphi^{-1}(y^s)$;

(6) Add the simulated value to the set of conditioning data;

(7) Loop until all the N nodes of the simulated grid have been simulated.

The resulting co-simulated models reproduce the marginal probability distribution of $F_{z2}(z)$, the experimental data values $(z_2(x_\alpha))$ at its own location, and the imposed spatial continuity pattern as well as the spatial correlation between both variables (Soares 2001).

3.4.3 Stochastic Sequential Co-simulation with Joint Probability Distributions

Independently of the joint probability distribution observed from the experimental data between the primary and secondary variables $(F(Z_2, Z_1))$, the simulated models resulting from the direct sequential co-simulation (described in previous point) are able only to reproduce a linear correlation between the primary and secondary variables (Fig. 3.4; Horta and Soares 2010). This is a clear limitation to the described co-simulation methodology.

In reservoir characterization, it is of extreme importance to create reservoir models capable of reproducing the joint probability distributions as estimated from the well log between the properties of interest. For example, in seismic inversion methodologies it is essential to reproduce the nonlinear relationships between density, P-wave and S-wave velocities (or between acoustic and elastic impedances) for facies classification over the simulated models. The reproduction of just a linear correlation between simulated properties is clearly insufficient for modeling complex subsurface petro-elastic models (Avseth et al. 2005).

The noted limitations of the traditional direct sequential co-simulation methodology were mitigated by the development of the direct sequential co-simulation with joint probability distributions (Horta and Soares 2010). Basically, this algorithm extends the concept of resampling global distributions to joint probability distributions in order to ensure the reproduction of the experimental joint probability distribution between the primary and secondary variables in the simulated models. In Fig. 3.4 we can see the bi-plot between Vp and density estimated from well-log data, the corresponding bi-plot between co-simulated models estimated using direct sequential co-simulation and the bi-plot between both variables obtained with direct sequential co-simulation with joint probability distributions.

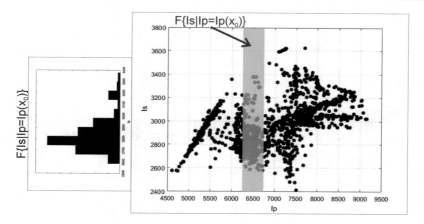

Fig. 3.5 Schematic representation of how the conditional cumulative distribution function, $F\{I_s | I_p = I_p(x_0)\}$, is estimated for the co-DSS with joint probability distributions

Following the sequential simulation approach, the direct sequential co-simulation with joint probability distributions procedure is based on the Bayes' rule. The simulation methodology assumes a previously simulated model, the secondary variable ($Z_1(x)$), is first calculated for the entire simulation grid with DSS, then the primary variable ($Z_2(x)$) is co-simulated.

For a given $Z_1^s(x_0)$ simulated at a node x_0 of a random path, the following conditional cumulative distribution function is estimated from the global bi-distribution (Fig. 3.5):

$$F\big(Z_2(x_0) | Z_1(x0) = z_1^s(x_0)\big). \qquad (3.9)$$

Local conditional mean and variance are estimated with collocated simple co-Kriging estimate (Eq. 3.7) and collocated simple co-Kriging variance (Eq. 3.8). Instead of resampling the global distribution $Z_2(x)$, the idea is to resample this conditional distribution (Eq.3.9) based on the local conditional mean and variance, to draw $Z_2^s(x_0)$.

Direct sequential co-simulation with joint probability distributions is able to reproduce the marginal probability distribution functions of the primary variable; the value of the experimental data is honored at its spatial locations and the spatial continuity model is also reproduced.

Moreover, the joint probability distributions, as estimated from the experimental data between the primary and secondary variables, $F(Z_2(x) | Z_1(x))$, are reproduced, which is the main purpose of this simulation algorithm.

This procedure can be summarized through the following sequence (Horta and Soares 2010):

(1) Estimate the global bi-distribution $F(Z_2(x), Z_1(x))$ from the experimental data;

(2) Simulate the secondary variable $Z_1(x)$ with DSS for the entire simulation grid;

(3) Following a random path, estimate the local mean and variance at x_0 with collocated simple co-Kriging estimate ($[Z_2(x_0)^*]_{CSK}$) and the corresponding co-Kriging variance ($\sigma_{CSK}^2(x_0)$) conditioned to the original experimental data and the previously simulated data ($z_2(x_\alpha)$) and the collocated datum of the secondary variable ($z_1^s(x_0)$);

(4) Based on the simulated values for the secondary variable, $z_1^s(x_0)$, calculate the conditional cumulative distribution function $F(Z_2(x) | Z_1(x) = z_1^s(x_0))$ from the global joint-probability distribution $F(Z_2(x), Z_1(x))$;

(5) Follow the traditional DSS approach to simulate the value for $z_2^s(x_0)$ from the conditional cumulative distribution function $F(Z_2(x) | Z_1(x) = z_1^s(x_0))$.

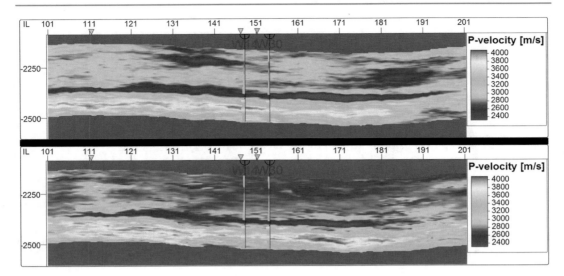

Fig. 3.6 Comparison between co-simulated models with direct sequential co-simulation (on the *top*) and with direct sequential co-simulation with joint probability distributions (on the *bottom*). The model shown in Fig. 3.3 was used as secondary variable for the co-simulation process. The joint probability distributions estimated from the experimental data and between primary and secondary variable for both co-simulation methods are shown in Fig. 3.4

The models resulting from the stochastic simulation with DSS with joint probability distributions ensure the spatial reproduction of the bi-distributions between the primary and secondary variables (Fig. 3.6).

3.4.4 Stochastic Simulation with Uncertain Data: DSS with Point Probability Distributions

The traditional DSS assumes that the available experimental data, like for example the existing well-log data, is considered hard-data with no uncertainty attached. However, in some situations there can be considerable uncertainty in the experimental well-log data. For example, there are cases where we can only access the upper and lower boundaries of pore pressure or porosity, and instabilities along the borehole cause uncertainty and wrong readings of the properties of interest. These uncertainties may be interpreted at the well sample data locations as probability distribution functions along the well path, instead of one hard value per sample. In these cases, the stochastic sequential simulation must account for these well-log data point distributions,[1] and also for the local distribution of the variable to be modelled as in the usual procedure (Sect. 3.4.1). To deal with point distribution, Srivastava (1992) and Froidevaux (1993) propose, using the P-field simulation, generating a spatial probability field to produce, using Monte Carlo, the RF of $Z(x)$. This method has presented some problems with variogram reproduction for three-dimensional high anisotropic fields, and also presents artefacts at the conditioning data location (Pyrcz and Deutsch 2002).

The proposed method aims at integrating the uncertainty of data measurements—local distributions—in the framework of a stochastic sequential simulation. It consists of a sequential approach where, in a first step and before simulating the entire grid, the experimental data values located at $x\beta$ are firstly drawn from the local

---

[1]The notation 'point distribution' is for the well-log data only.

distributions, and in a second step the entire grid of nodes is sequentially simulated, conditioned to the previously simulated experimental data. The data values $z(x_\beta)$ drawn from $F(z(x_\beta))$ must reproduce not only these local cfds, but also the spatial correlation as revealed by the spatial covariance.

The idea of generating a spatially correlated set of data values consists of two steps: calculating the local mean and variance at the experimental data location, conditioned to the neighborhood "hard" data $z(x_\alpha)$ and also to the values previously drawn from other uncertain data distributions; and, afterwards, generating the simulated values with the local distributions, centered at the local mean and variance, following the outline of direct sequential simulation (Soares 2001).

The simulated data values, as they come from distributions centered at simple kriging mean and variance, reproduce not only the spatial covariances (Journel 1994), but also the local distributions (Soares 2001). This is a development of a version of direct sequential simulation with joint probability distributions (Horta and Soares 2010), which can be summarized into two basic steps:

1. The direct sequential simulation starts by generating first the N_d values at the experimental sample data locations, $x_\beta = 1, \ldots, N_d$ using the local distributions $F(z(x_\beta))$. At a given sample location x_β, the mean and variance of $z(x_\beta)$ are calculated by simple kriging based on the known "hard" data $z(x_\alpha)$ and previously simulated uncertain experimental data $z^s(x_\beta)$. A value $z^s(x_0)$ at the experimental location x_0 is drawn from the local distribution $F(z(x_0))$ centered at the simple kriging estimate of local mean:

$$z^s(x_0) = \sum_\lambda \lambda_\alpha(x_0)z(x_\alpha) + \sum_\beta \lambda_\beta(x_0)z^s(x_\beta),$$

and with a local variance identified with simple kriging variance where λ_α and λ_β are, respectively, the kriging weights associated with the known "hard" data and the previously simulated uncertain experimental data.

2. After a set of N_d experimental sample data is generated from the local distributions, in a second stage, the direct sequential simulation methodology generates $z(x)$ values on the entire grid of nodes conditioned to the simulated values at the experimental locations plus the eventual hard data $\{z(x_\alpha), \alpha = 1, \ldots, N; z(x_\beta), \beta = 1, \ldots, N_d\}$. For each realization, a new set of sample data is generated before the rest of the model is simulated.

3.5 Simulation of Categorical Variables

3.5.1 Indicator Simulation

A common practice in hydrocarbon reservoir characterization consists of describing petrophysical properties by using the concept of facies (lithofacies, lithogroups, rocktypes) to define lithologies with statistical homogeneity of the property of interest (e.g. high porosity, low permeability, etc.). The aim is to characterize the spatial dispersion of those homogeneous facies instead of the spatial dispersion of the petrophysical properties themselves. These facies can thus be considered as categorical variables.

This section will present a set of categorical variable simulation methodologies that describe the shape of bodies in spatial phenomena. The aim of these simulation methodologies is to generate indicator simulation models with the same statistical characteristics in terms of dispersion and spatial continuity as the set of available sample values. These algorithms are based on an estimation of categorical variables presented in Chap. 2.

For the simulation of a set of facies X_k, $k = 1$, ..., N, which are mutually disjunctive, the sequential indicator simulation algorithm, which is based on the estimation methods of categorical variables presented in Chap. 2, can be described as follows:

(1) Define a vector with the categorical variables $I_k(x)$:

$$I_k(x) \begin{cases} 1 & if \ \ x \in X_k, \quad k = 1, \ldots, K \\ 0 & if \ \ x \in X_j, \quad j \neq k. \end{cases} \qquad (3.10)$$

For each phase X_k we may define the mean, $m_k = E\{I_k(x)\}$, which is the proportion of X_k in A, where the spatial continuity of the group $I_k(x)$, $k = 1$, ..., K, may be measured by the multiphase covariance (Eq. 3.11) or variogram (Eq. 3.12):

$$C(\boldsymbol{h}) = E\left\{ \sum_{k=1}^{K} [I_k(x) \cdot I_k(x+\boldsymbol{h})] \right\}, \qquad (3.11)$$

$$\gamma(\boldsymbol{h}) = \frac{1}{2}E\left\{ \sum_{k=1}^{K} [I_k(x) - I_k(x+\boldsymbol{h})]^2 \right\}. \qquad (3.12)$$

(2) Calculate the Kriging estimate of the probabilities for a given location x_1 randomly selected within the study area A,:

$$[I_k(x_1)]^* = prob\{x_1 \in X_k\}^*, \quad k = 1, \ldots, K. \qquad (3.13)$$

The estimated values $[I_k(x_1)]^*$, $k = 1$, ..., K may be obtained by taking the individual covariance of each single phase or by the multi-phase covariance into account. The latter

implies the resolution of just one single Kriging system.

(3) Correct the eventual violation of the order relations:

The sum f the estimated local probabilities values must be one:

$$\sum_{k=1}^{K} [I_k(x_1)]^* = 1. \qquad (3.14)$$

The multi-phase Kriging ensures this relation since it uses a single model to estimate the K values $[I_k(x_1)]^*$, $k = 1, \ldots, K$.

However, it may be that the estimated values are not comprised within the interval $[0,1]$ —$I_k(x_1)^* > 1$ or $I_k(x_1)^* < 0$ resulting from negative weights calculated from the individual or multi-phase Kriging system. In these situations, we need to make corrections for the order relations.

(4) Compute a 'pseudo' cumulative function:

$$[F_l(x_1)]^* = \sum_{k=1}^{K} I_k(x_1)^*, \quad k = 1, \ldots, K. \qquad (3.15)$$

(5) Generate a p value uniformly distributed between 0 and 1. So:

x_1 belongs   to   phase k if $p \in [[F_{k-1}(x_1)^*], [F_{kl}(x_1)]^*]$ i.e. the value simulated in x_i is transformed into: $I_k(x_1) = 1$ and $I_j(x_1) = 0$, $j \neq k$.

6) The simulated values $I_k(x_1)$, $k = 1$, ..., K are integrated into the group of conditioning data for the simulation of the next location. The sequence from (1) to (6) is repeated until all locations in A are visited.

There are cases in which the sequential indicator simulation introduces a bias in the final proportions of facies and consequently does not reproduce the global proportion of each phase: the final proportions of each phase are dependent on the location of the first points randomly simulated within *A*. If the first simulated points are located near the available experimental data of a given phase, the proportion of estimated values for that specific phase tends to increase rapidly during the sequential simulation procedure and will hardly reproduce the global proportions as estimated from the experimental data. This effect can be severe if the variogram ranges are large, affecting mainly the classes of smaller proportions. Soares (1988) proposed a method for the correction of local probabilities according to the global proportions of facies during the sequential process.

3.5.2 Alternative Simulation Methods for Categorical Variables

Alternatively, there is a class of sequential simulation methodologies that use the sequential Gaussian simulation (see Sect. 3.3) to simulate categorical variables: for example, truncated Gaussian or pluri-Gaussian simulation. The idea is to generate one or several Gaussian random fields using the sequential Gaussian simulation as described in Sect. 3.3 and categorize classes of one Gaussian distribution with thresholds identified with the proportions of different facies (truncated Gaussian) or categorize the facies with a joint truncation of two or pluri-Gaussian fields. In this case, the thresholds are calculated based on a model of a joint distribution of facies (Galli et al. 1994; Le Loc'h et al. 1994).

3.5.3 High-Order Stochastic Simulation of Categorical Variables

Categorical variables can also be simulated based on multi-point statistics (Chap. 2). Multi-point statistics overcome the limitations of two-point statistics to reproduce connectivity patterns by simultaneously using a set of points for the inference of the property of interest at a given location (Iaco and Maggio 2011). Within this high-order statistics framework the following references summarize the best known multi-point geostatistics techniques for modeling in Earth sciences: Strebelle (2002), Arpat and Caers (2007), Daly and Caers (2010), Mariethoz et al. (2010), Mariethoz and Caers (2014). More recently, the use of high-order statistics, or cumulants, have been proposed for modeling non-Gaussian Earth phenomena (Dimitrakopoulos et al. 2010).

Multi-point geostatistics is based on the concept of training images, which are defined as a conceptual representation of reality. The use of training images overcomes the lack of experimental data for retrieving reliable high-order statistics: they are inferred directly from the training image. Thus, the training image must represent all expected structures for a given study area as well as the spatial distribution of the property of interest. In reservoir modeling, this training image is often provided by a geologist who has a general idea of the structural and sedimentary subsurface geology of the area being studied. The best known multi-point geostatistical algorithms and software used for reservoir modeling are SNESIM (Strebelle 2002), SIMPAT (Arpat and Caers 2007) and direct sampling (Mariethoz et al. 2010).

Integration of Geophysical Data for Reservoir Modeling and Characterization

4

The second part of this book deals with the integration of different types of information for characterizing hydrocarbon reservoirs with high spatial resolution models of petro-elastic properties of interest: e.g. lithologies, facies and fluid saturations. These modeling techniques are based on the geostatistical methods presented above and their advantages are related to the integration of data with different support (time and spatial support) and uncertainty, such as seismic reflection and well data, in coherent and realistic models of hydrocarbon reservoirs.

Traditionally, reservoir models were built exclusively with information retrieved from sparsely located wells, and possibly conditioned to a secondary variable of interest (e.g. a trend map or geological model) provided by an expert who was usually a geologist. Although well data provides certain 'hard' measures of the subsurface property of interest, in most cases—given the low number of available wells—the data lacks of spatial representativeness and, consequently, the corresponding models provide little understanding of the complex and variable subsurface geology of the entire reservoir area (Dubrule 2003).

Since it has high spatial representativeness by covering the full spatial extent of the reservoir volume, seismic reflection data is a different window for the subsurface properties of interest. However, seismic reflection data has a poor spatial resolution along the vertical direction (temporal domain), which means much greater support compared with the well-log data and much greater uncertainty derived both from measurement errors and the nonlinear relationship between the recorded seismic signal and the subsurface properties we wish to describe.

Given the rich spatial extent of seismic reflection data, the use of such information as conditioning data for the three-dimensional modeling of hydrocarbon reservoirs has been the purpose of geostatistical methods based on direct approaches. These include simulation using the seismic reflection data either as a trend, or joint simulation with seismic reflection data (or any other seismic attribute, such as acoustic impedance) as a secondary variable (Dubrule 2003; Doyen 2007). The problem with these approaches are the simultaneous use of different support data: i.e. different spatial or temporal scales, and the usually poor relationship between seismic amplitudes and the subsurface petro-elastic properties of interest.

As for the integration of different support data, Liu and Journel (2009) propose a new DSS method—direct block sequential simulation—for the integration of coarse (seismic data) and fine scale (well-log data) data. But this lacks a relationship between seismic amplitudes and the subsurface petrophysical properties—a necessary condition for any of the usual methods of simulation with trends or joint simulation. This remains a major limitation for these approaches.

However, if the seismic reflection data or its amplitudes is usually poorly related to the subsurface petrophysical properties (such as facies, porosity and saturation), other seismic attributes,

© Springer International Publishing AG 2017
L. Azevedo and A. Soares, *Geostatistical Methods for Reservoir Geophysics*,
Advances in Oil and Gas Exploration & Production, DOI 10.1007/978-3-319-53201-1_4

such as acoustic and/or elastic impedances, can have a significant relationship with some facies and/or classes of porosity. As these parameters (e.g. acoustic/elastic impedances) are physically related to seismic amplitudes, this induced a new class of methods that are based on the inverse solution of a simple, but ill-posed problem and with non-unique solutions: on wishes to know the model parameters (reflectivity coefficients derived from the subsurface elastic properties), which convolved with a known wavelet originate of the known solution (i.e. the recorded seismic amplitudes).

The theoretical solutions for seismic inversion are stated in Tarantola (2005), while a brief introduction to seismic inversion can be found below (Sect. 4.1). The seismic inversion problem began to be tackled with deterministic methodologies, which are optimization procedures seeking the minimization of an objective function—normally the mismatch between the synthetic seismic that is obtained by perturbing an initial guess and the observed seismic reflection data (Lindseth 1979; Lancaster and Whitcombe 2000; Russell 1988; Coléou et al. 2005).

In recent decades, seismic inversion has been successfully extended to a statistical framework for assessing the uncertainty of the inferred 3D subsurface elastic models, which is one of the major limitations of deterministic inverse procedures. Among the many statistical inverse approaches, two different stochastic approaches for solving the seismic inversion are worth mentioning.

The first group of stochastic methodologies approaches the seismic inversion as an optimization problem in an iterative and convergent process. This includes what are traditionally designated iterative geostatistical seismic inversion methods, introduced in the seminal work by Bortolli et al. (1993). The latter authors use a stochastic sequential simulation algorithm (sequential Gaussian simulation, Sect. 3.3) to generate a spatial RF of acoustic impedance traces. The optimization process is made trace-by-trace, which became the main limitation of the method due to its inability to distinguish between signal and noise. Alternative methodologies that

overcame these limitations were presented later (e.g. Soares et al. 2007).

The second group of stochastic seismic inversion algorithms is called linearized Bayesian inverse methodologies. These are based on a particular solution of the inverse problem using the Bayesian framework and assume the parameters and observations are multi-Gaussian distributed as well as the data error, which allows the forward model to be linearized (Buland and Omre 2002). Several authors have recently contributed towards overcoming some of the limitations of this method, particularly the multi-Gaussian assumption, by using Gaussian Mixture Models (Grana and Della Rossa 2010).

The seismic inversion methodologies this book focuses on are those based on geostatistical iterative procedures, in which the model parameter space is globally perturbed by stochastic sequential simulation algorithms. These geostatistical inverse methods are based on the family of DSS and joint co-simulations, with local and point distributions (Chap. 3). As they do not imply any nonlinear transformation of parameters or observations, these methods have a high potential for accommodating accurate solutions for new challenges of different data integration, such as the joint inversion of seismic reflection and electromagnetic data or the integration of production data into seismic inversion (Chap. 6).

4.1 Seismic Inversion

Predictions about an Earth's physical system (e.g. the Earth's subsurface petro-elastic properties) can be made by assuming a theoretical model that globally explains the system being studied. The process of forecasting a response for a particular physical system (e.g. weather forecasting) is commonly called forward modeling. In forward processes we try to model the parameters of the system to obtain its solution. On the other hand, in inverse physical problems, such as the seismic inversion problem, we know the Earth's response to a limited set of indirect measurements and we try to infer data about the model parameters of the system in studies that give rise to that solution (Tarantola 2005).

Geophysical inverse problems aim to infer the physical properties of the subsurface geology, the model parameters ($\mathbf{m} \in \mathbf{R^n}$), from a set of indirect geophysical measurements/observations ($\mathbf{d_{obs}} \in \mathbf{R^n}$) that are normally contaminated by measurement errors ($\mathbf{e}$) originating from different sources. The observed data ($\mathbf{d_{obs}}$) and the subsurface properties of interest ($\mathbf{m}$) are connected by a forward model ($\mathbf{F}$). If the forward models can be mathematically described and the model parameters are known, the observed data may be synthesized by Eq. 4.1 (Tarantola 2005):

$$\mathbf{d_{obs}} = \mathbf{F(m)} + \mathbf{e}. \tag{4.1}$$

In the particular case of seismic inversion problems, $\mathbf{d_{obs}}$ represents the recorded seismic reflection data and well-log data (if available), $\mathbf{F}$, is normally defined as the convolution model and $\mathbf{m}$ the model parameter space for the properties to invert. These properties depend on the goal of the inversion: acoustic and/or elastic impedances or density, P-wave and S-wave velocity models. The forward model, $\mathbf{F}$, of Eq. 4.1 can be described by, for example, Eq. 1.1.

Seismic inversion problems are nonlinear, ill-conditioned and with non-unique solutions due to the intrinsic limitations of the seismic method itself: limited bandwidth and resolution of the seismic reflection data, noise, measurement errors, numerical approximations and physical assumptions about the involved forward models (Tarantola 2005; Bosch et al. 2010; Tompkins et al. 2011). Assuming the forward model is valid, the optimal inverse elastic models retrieved at the end of a seismic inverse process are just one set of possibilities among several Earth models that equally satisfy the observed seismic reflection data. Due to the non-unique solution we may say that if the match between the recorded seismic data and the synthetic seismic reflection data, calculated from the best-fit inverse models, is poor, then we can conclude the correspondence between the real elastic models with the inverted ones is also poor. However, the opposite may not be true. One can achieve a good match between

the observed and inverted synthetic seismic reflection data while the real and inverted subsurface models are not converging with each other—the inverse solution is converging towards a local minimum far from the global minimum (Tarantola 2005). The non-unique nature of seismic inversion is a critical aspect of the methodology approach. Regardless of the chosen methodology and the underlying assumptions for solving the seismic inverse problem, there is always uncertainty with the inverted elastic models that needs to be continuously assessed and propagated during the inversion procedure (Bosch et al. 2010; Grana and Della Rossa 2010; Tompkins et al. 2011).

Inverted elastic subsurface Earth models are now routinely used in reservoir modeling and characterization studies. Therefore, it is important to understand the different seismic inverse methodologies available and the underlying assumptions associated with each (e.g. assumptions about prior probability distributions and about the spatial continuity patterns). These assumptions have a significant impact on the exploration of the model parameter space and, consequently, on the assessment of the uncertainty of the best-fit inverse models. Seismic inversion problems can be divided into two main approaches: the deterministic (also called optimization techniques) and probabilistic (Bosch et al. 2010).

Band-limited, or integration of the seismic trace (Lindseth 1979), colored inversion (Lancaster and Whitcombe 2000) and sparse-spike and model-based (Russel 1988) are the main deterministic algorithms for post-stack seismic inversion. Of this group of inversion procedures, the sparse-spike and model-based methodologies are the most widespread deterministic inversion techniques among the geophysical community (Bosch et al. 2010). Sparse-spike inversion is a model-driven technique that tries to sparsely estimate the real reflection coefficients along the seismic trace by deconvolution. With this approach, a minimum number of reflections are considered in reproducing the real seismic trace after being convolved with a wavelet. The sparse inverted impedances are then combined with a low-frequency model in order to better constrain the observed data and add spatial consistency to the inverted traces (Russel 1988; Russell and Hampson 1991; Bosch et al. 2010). In the model-based approach, an initial subsurface model, which is normally designated as a low frequency model, is applied to the inversion algorithm and then perturbed until it produces synthetic seismic correlated enough with the recorded seismic (Russel 1988; Russell and Hampson 1991). Within these frameworks, the inverted impedance models provide a smooth representation of the Earth's subsurface properties with much less spatial variability compared to the real and complex subsurface petrophysical and geological properties (Russell and Hampson 1991).

Beside the widespread use of these methodologies within the industry, the uncertainty assessment of deterministic solutions is limited. Within this framework the uncertainty can only be assessed by a linearization around the best-fit inverse solution, which is normally retrieved by least squares. Due to this limitation, these methodologies are not suitable for highly nonlinear inverse problems such as pre-stack seismic inversion and complex geological environments (Tarantola 2005; Tompkins et al. 2011).

Among the stochastic methods, Bayesian approaches ensure the propagation of the uncertainty from the prior probability distributions estimated from the experimental data (e.g. well-log data), to the probability distributions of the model parameters space (Grana et al. 2012). Within this framework, the linearity assumptions of the deterministic solutions are overcome, allowing for a more comprehensive exploration of the uncertainty and the model parameter space. However, the uncertainty assessment depends on the parameterization of the inverse problem—e.g. assumptions about the prior distributions and the spatial continuity pattern (Scales and Tenorio 2001).

Geostatistical approaches define the inverse solution as a probability density function on the model parameters space (Bosch et al. 2010). Normally, the inverse solution is achieved by sampling the model parameter space by Monte Carlo or by using geostatistical sequential simulation combined with global optimization algorithms. Genetic algorithms (e.g. Mallick 1995, 1999; Boschetti et al. 1996; Soares et al. 2007) and simulated annealing (Sen and Stoffa 1991; Ma 2002) fall within this class of inverse methodologies. When compared with deterministic inverse methodologies, these stochastic approaches are normally much more computationally expensive (Bosch et al. 2010).

4.2 Bayesian Framework for Integrating Seismic Reflection Data into Subsurface Earth Models

Methodologies commonly designated as linearized Bayesian inversion (e.g. Buland and Omre 2003; Buland and El Ouair 2006; Grana and Della Rossa 2010) assume the linearization of the forward operator, and multi-Gaussian distribution for the prior probability distributions and for the errors within the observed seismic data. The resulting inverse solutions are mathematically tractable and, due to the referred assumptions, the computational time is much lower when compared with iterative geostatistical approaches based on genetic algorithms (Sect. 4.3) or simulated annealing. However, the reduction of the computational burden has a direct impact on the exploration of the uncertainty space even in the presence of exact prior information (Scales and Tenorio 2001; Tarantola 2005; Bosch et al. 2010). To overcome the limitations imposed by the Gaussian assumption of the prior probability distributions while maintaining the computational efficiency of the Bayesian linearized procedures, Grana and Della Rossa (2010) developed an inversion framework for the Bayesian linearized inversion using Gaussian-mixture models. Within their method, and by defining the prior probability distributions as Gaussian-mixtures and using a linearized forward model, the inferred posterior probability distribution also is analytically expressed as Gaussian-mixtures.

Due to the importance of Bayesian inversion methodologies (Tarantola 2005; Buland and Omre 2003; Buland and El Ouair 2006) in seismic reservoir characterization, here we introduce the main principles shared by these linearized inverse methodologies.

Let us assume the model parameters space is defined by $\mathbf{m}$, e.g. subsurface P-wave velocity, S-wave velocity and density, and the observation data, $\mathbf{d}$, the recorded seismic reflection and the well data. The solution of a Bayesian inverse problem is represented by the posterior distribution $p(\mathbf{m}|\mathbf{d})$, as the product of a likelihood function $p(\mathbf{d}|\mathbf{m})$ and an a priori model $p(\mathbf{m})$:

$$p(\mathbf{m}, \mathbf{d}) = p(\mathbf{d}|\mathbf{m}) \cdot p(\mathbf{m}). \qquad (4.2)$$

In Bayesian seismic inversion, the vectors $\mathbf{m}$ and $\mathbf{d}$ are assumed to be multi-Gaussian distributed (Buland and Omre 2003; Grana and Della Rossa 2010). Hence, the posterior conditional distribution $p(\mathbf{m}|\mathbf{d})$ is a multi-Gaussian distribution $N(\mu_{m|d}, \sum_{m|d})$ with conditional mean $(\mu_{m|d})$ and covariance $(\sum_{m|d};$ Anderson 1984):

$$p(\mathbf{m}|\mathbf{d}) = N\left(\mu_{m|d}, \sum_{m|d}\right), \qquad (4.3)$$

where $\mu_{m|d} = \mu_{m|d} + \sum_{m|d} \sum_{d}^{-1} (\mathbf{d} - \mathbf{m})$ and $\sum_{m|d} = \sum_m - \sum_{m|d} \sum_d^{-1} \sum_{m|d}$, where Σ are the covariance matrixes of $\mathbf{m}$, $\mathbf{d}$ and cross-covariance $\mathbf{m}$, $\mathbf{d}$.

This is the particular case, when looking for the solution in the multi-Gaussian space, of the more general inverse problem posed by Tarantola (2005).

Even in a linearized version of the solution (e.g. Buland and Omre 2003; Buland and El Ouair 2006), in which the convolution of the wavelet and reflectivity coefficients are written as a linear operator, Eq. 4.2 cannot be directly analytically calculated. Hence, the stochastic model proposed by some authors (e.g. Buland and Omre 2003; Buland and El Ouair 2006) entails the resampling of the posterior distribution with a Markov Chain Monte Carlo (MCMC) algorithm, often the Metropolis-Hastings algorithm (or in a more particular case, the Gibbs sampling).

In Bayesian inversion, MCMC sampling consists of drawing spatial RFs of the elastic properties of interest (e.g. P-wave and S-wave velocities and density) and, at each iteration, checking the acceptability of the new realization through the likelihood function. The a posteriori resampling method must assure the convergence to stable solutions: i.e. the a posteriori distributions. Hence, the role of the a priori model is extremely important for guaranteeing the stability of the convergence process. The a priori model $p(\mathbf{m})$ is normally chosen to contain the low frequencies missing from the available seismic reflection data. The low frequency model can be a Krigged model for the area of interest using the available well-log data as experimental data.

In spite of their implementation simplicity, Bayesian inversion methodologies have some limitations. The main one is related to the multi-Gaussian assumption between all parameters and the observed data. Some authors argue the process assumes multi-Gaussian just for the residuals around a given a priori model $p(\mathbf{m})$. In this case, the a priori model assumes a key role in the reliability of the inverse elastic model. The second limitation is the temporal resolution of the resulting inverse models, which is limited to the resolution of the seismic data.

The limitations imposed by the Bayesian framework may be overcome by using inversion methodologies that are able to account for non-Gaussian and the non-linearization of the forward operator. The most common inversion techniques matching these criteria are iterative procedures based on stochastic sequential simulation.

4.3 Iterative Geostatistical Seismic Inversion Methodologies

The aim of a seismic inversion study is the inference of the subsurface elastic or acoustic properties from recorded seismic reflection data. Depending on the limitations associated with the chosen inversion methodology, the retrieved inverse models can be acoustic and/or elastic impedance for post-stack seismic data, or density, P-wave and S-wave models if a more elaborate inversion algorithm is being used to invert pre-stack seismic reflection data (Francis 2006).

Seismic inversion is an ill-posed, nonlinear problem with multiple solutions that can be summarized by Eq. 4.1 (Tarantola 2005). The goal is to estimate a subsurface Earth model, **m**, that after being forward modelled, **F**, produces synthetic seismic data showing a good correlation with the recorded seismic data. The match between observed and synthetic seismic is achieved by the maximization (or minimization) of an objective function measuring the mismatch between inverted and real seismic. For example, the objective function can be the Pearson's correlation coefficient (Eq. 4.4):

$$\rho_{X,Y} = \frac{cov(X, Y)}{\sigma_X \sigma_Y}, \qquad (4.4)$$

where *cov* is defined as the centered covariance between variables X and Y, which are the synthetic and real seismic volumes, respectively, and σ the individual variances of each variable. More complex objective functions integrate Pearson's correlation coefficient with least-square errors calculated between the synthetic and the recorded seismic reflection data in terms of amplitudes.

Due to the intrinsic nature of the seismic inversion problem, it is appropriate to pose it in a statistical framework, such as the geostatistical model introduced in Chap. 2. Stochastic inverse solutions are able to provide more realistic and heterogeneous models when compared with those retrieved from deterministic or linearized approaches. It is important to reproduce the main aspects of the geological, petrophysical and rock

physics components of the reality in the inverted elastic models, since they are fundamental for reserve calculations and fluid flow simulations (Francis 2006; Bosch et al. 2010).[1]

A geostatistical seismic inversion framework consists of an iterative procedure in which a set of realizations of parameters, **m**, are generated by the stochastic sequential simulation methods described in Chap. 3 and optimized until the match of the objective function reaches a given user-defined value.

By using the geostatistical methodologies presented in the previous chapter we seek to reproduce the main spatial continuity patterns of elastic and acoustic properties, and petrophysical properties models in the solution of Eq. 4.1, and to access the uncertainty attached to those models.

4.3.1 Frequency Domain of Geostatistical Seismic Inversion

One of the main characteristics of geostatistical seismic inversion is producing high-resolution subsurface Earth models. The high temporal frequencies are not directly inverted from the seismic data since they are not part of the recorded seismic data due to its band-limited nature. Hence, the small-scale variability in geostatistical inverse models is related to the much higher frequency content present in these inverse models compared with those retrieved from determinist solutions (Fig. 4.1). The integration of the well-log data, also with a very high vertical resolution, is another source for the high-frequency content models retrieved from geostatistical seismic inversion methodologies. Unlike the deterministic approaches, in which the

---

[1]It is important to stress that the choice of inverse methodology should take into account the goal of the study versus the computational effort involved in the inversion procedure and the quality of the available seismic data. Often a cheap solution, such as a deterministic one, compared to a stochastic approach, may be enough for the identification and delineation of the main reservoir areas in early exploration stages.

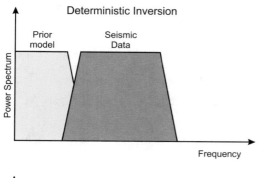

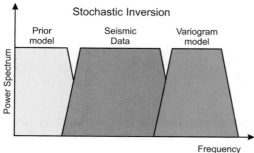

Fig. 4.1 Schematic representation of the comparison between the bandwidth extension of inverted models from deterministic and stochastic solutions. The high temporal frequency content of the stochastic solution is provided by the spatial continuity model imposed for the stochastic simulation (adapted from Dubrule 2003)

well-log data is only considered for estimating an initial low-frequency model, the integration of well-log data into geostatistical seismic inversion methodologies allows the well-log data to be honored at its location (Doyen 2007; Filippova et al. 2011).

However, because of the very different frequency content of the seismic and well-log data, the integration of the well-log into the inversion grid is not straightforward. While the maximum frequency content of standard seismic reflection data is around 75 Hz, the frequency content of the well-log data is in the order of hundreds of Hertz. The differences in frequency content are translated into different data sampling rates. In order to ensure the same sampling rate, it is a common approach to upscale the well-log data into the seismic scale.

There are many different upscaling methodologies, depending on the assumptions made. It is important to know and understand the

assumptions behind the upscaling technique selected, since it can dramatically change the content of the original well-log data. In statistical frameworks it is advisable to select an upscaling methodology that reproduces original first and second order statistical moments, the mean and variance, as estimated from the available well-log data.

Sometimes the low-frequency bandwidth is filled-in by a low-frequency model computed pre-inversion (see Sect. 4.4). This model is used as constraining data for the inversion procedure and is frequently created by Kriging the well-log data of the properties of interest after they are filtered for the frequencies of interest (normally between 0 and 10 Hz).

4.3.2 Trace-by-Trace Geostatistical Seismic Inversion Methodologies

The first geostatistical seismic inversion methodology was introduced by Bortoli et al. (1992) and Haas and Dubrule (1994). These authors proposed a sequential trace-by-trace approach (Fig. 4.2) in which each seismic trace, or CMP location, is visited individually following a pre-defined random path within the seismic volume. At each step along the random path a set of Ns realizations of one acoustic impedance trace is simulated using SGS (Sect. 3.3), taking the well-log data and previously visited/simulated nodes into account. Then, for each individual simulated impedance trace, the corresponding reflection coefficient is derived and convolved by a wavelet, resulting in a set of Ns synthetic seismic traces. Each of the Ns synthetic traces is compared in terms of a mismatch function with the recorded/real seismic trace. The acoustic impedance realization that produces the best match between the real and the synthetic seismic traces is retained in the reservoir grid as conditioning data for the simulation of the next acoustic impedance trace at the new location following the pre-defined random path (Bortoli et al. 1992; Haas and Dubrule 1994).

As with any stochastic sequential simulation approach, since the random path changes on each

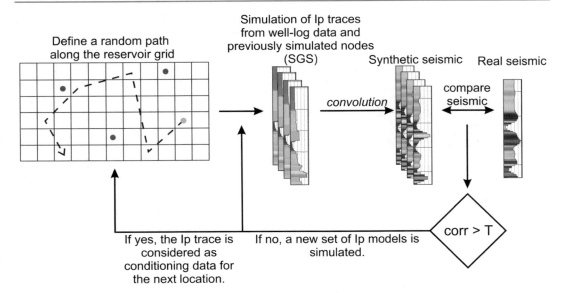

Fig. 4.2 Schematic representation of trace-by-trace geo-statistical inverse methodologies. At the first location (*blue circle*) a set of *Ns* acoustic impedance traces is simulated from the available well-log data (*red circles*). Each simulated trace is convolved with an estimated wavelet producing a set of *Ns* synthetic seismic traces, which are compared with the real seismic trace at that location. If the correlation coefficient (corr) is above a certain threshold (T), the Ip trace is considered to be conditioning data for the location along the random path: otherwise, a new set of Ip traces is simulated

individual geostatistical inversion run—consequently modifying the conditioning data at each trace location—different runs produce variable inverted acoustic impedance models that fit the observed seismic reflection data equally. The variability among a set of inverted models allows the assessment of the spatial uncertainty related with the property of interest.

One of the main drawbacks of trace-by-trace stochastic seismic inversion methodologies concerns those areas of the recorded seismic reflection data with low signal-to-noise ratio. In areas of poor seismic signal, the sequential trace-by-trace approaches impose inverted models fitting the observed noisy seismic reflection data. As the simulated trace is assumed to be 'real' data for subsequent steps, this can lead to the spread of noisy values. In fact, the retrieved inverse impedance models should not fit the noise component of the signal, but instead should only fit the component of the signal corresponding to real subsurface geology. Noisy areas should be interpreted as high uncertainty areas with very low influence throughout the inversion process.

More recent versions of trace-by-trace models try to overcome this drawback by avoiding noisy areas in the early stages of the inversion, jumping to the trace location where the simulated Ip trace does not produce very high correlation coefficients compared with the recorded seismic, and revisiting these locations later in the inversion procedure. In this way, there is a larger degree of conditioning data producing a good match with the real seismic (Grijalba-Cuenca and Torres-Verdín 2000).

4.3.3   Global Geostatistical Seismic Inversion Methodologies

To overcome these limitations, Soares et al. (2007) introduce the global stochastic inversion methodology that, contrary to trace-by-trace approaches, uses a global approach during the stochastic sequential simulation stage. The general outline of this new family of geostatistical inversion algorithms is synthetized in Fig. 4.3. It is an iterative inverse approach that uses the

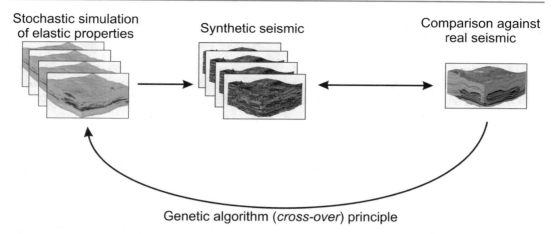

Genetic algorithm (*cross-over*) principle

Fig. 4.3 General outline for iterative geostatistical seismic inversion methodologies with a global approach. A set of *Ns* elastic models is simulated. *Ns* synthetic seismic volumes are derived from the simulated models. The synthetic seismic cubes are then compared against the real one in a trace-by-trace basis. The elastic traces that produce the synthetic seismic with a higher match against the real one are used as seed for the generation of a new set of models for the next iteration

principle of cross-over genetic algorithms as the global optimization technique, while the model perturbation towards the objective function is performed using direct sequential simulation and co-simulation (Soares et al. 2007; Caetano 2009).

At each iteration a set of *Ns* impedance models of the entire reservoir grid is generated. This geostatistical inversion approach is based on two key ideas: (i) the use of the direct sequential simulation and co-simulation as the method of perturbing the 3D impedance models in an iterative process; and (ii) to follow the sequential procedure of the genetic algorithms optimization to converge the transformed models towards an objective function: the global and local correlation coefficients between the transformed traces and the real seismic traces. These correlation coefficients of different simulated models are used as the affinity criterion between real and inverted seismic reflection data to create the next generation of models. The iterative procedure continues until a stopping criteria is reached: frequently the global correlation coefficient between real and inverted seismic reflection data.

With this approach, the areas of low signal-to-noise ratio remain poorly matched throughout the inversion procedure: an ensemble of best-fit inverted models will always present high variability or high uncertainty for those noisy areas where the signal-to-noise ratio is low.

This approach was generalized for the inversion of elastic properties, direct inversion of petrophysical properties and seismic AVA inversion, giving rise to the group of methodologies presented in this chapter.

The use of Direct Sequential Co-simulation for Global Transformation of Subsurface Earth Models

The use of direct sequential co-simulation as the model parameter space perturbation during the iterative geostatistical seismic inversion procedure is a key concept in all inverse methodologies presented in Sect. 4.3. The theoretical background behind this approach can be described as follows.

Let us consider one wishes to obtain a transformed model $Z_t(x)$ based on a set of N_i models $Z_1(x), Z_2(x), \ldots, Z_{Ni}(x)$ with the same spatial dispersion of the first and second order statistics, e.g. covariance and variogram ($C_1(\mathbf{h})$, $\gamma_1(\mathbf{h})$) and global histogram ($F_{zt}(z)$). We may generate $Z_t(x)$ by direct co-simulation, having $Z_1(x), Z_2(x), \ldots, Z_{Ni}(x)$ as auxiliary variables (Chap. 3). The collocated co-Kriging estimator of $Z_t(x)$ can be generalized for N_i collocated variables by:

$$Z_t(x_0)^* - m(x_0) = \sum_\alpha \lambda_\alpha(x_0)[Z_t(x_\alpha) - m(x_\alpha)] + \sum_{i=1}^{N_i} \lambda_i(x_0)[Z_i(x_0) - m(x_0)]. \tag{4.5}$$

Since the models $\gamma_i(\boldsymbol{h}), i = 1, \ldots, Ni$ and $\gamma_t(\boldsymbol{h})$ are the same, the application of the Markov approximation (Almeida and Journel 1994) is quite appropriate: the corregionalization models are entirely defined with the correlation coefficients $\rho_{t,i}(0)$ between $Z_t(x)$ and $Zi(x)$. Therefore, the affinity of the transformed model $Z_t(x)$ with the multiple models $Z_i(x)$ is determined by the correlation coefficients $\rho_{t,i}(0)$. Hence, one can select the models with the characteristics we wish to preserve in the transformed model $Z_t(x)$.

Local Screening Effect of Conditioning Data

Let us assume that to estimate $Z_t(x_0)$, the collocated value $Z_i(x_0)$ of a specific model $Z_i(x)$, with the highest correlation coefficient $\rho_{t,i}(0)$, screens out the influence of the remaining collocated values $Z_j(x_0)$, $j \neq i$. Hence, we may estimate $Z_t(x_0)$ with just one auxiliary variable (the collocated data with the highest correlation coefficient at location x_0):

(Chap. 3), the joint transformation/simulation of elastic properties into the petrophysical property of interest.

By using collocated co-Kriging we ensure these auxiliary models will either have a strong or a limited influence when generating the new models, depending on the local match between the synthetic and real seismic reflection data found in previous iterations. On the other hand, the spatial continuity models and marginal and joint probability distributions as retrieved from the available well-log data will be honored on each single impedance realization.

4.3.4 Global Geostatistical Acoustic Inversion

The global stochastic inversion (GSI; Soares et al. 2007; Caetano 2009) allows the inversion of post-stack seismic reflection data for acoustic impedance (Ip) models.

$$Z_t(x_0)^* - m(x_0) = \sum_\alpha \lambda_\alpha(x_0)[Z_t(x_\alpha) - m(x_\alpha)] + \lambda_i(x_0)[Z_i(x_0) - m(x_0)]. \tag{4.6}$$

With the local screening effect, N_i models $Z_i(x)$ give rise to just one auxiliary variable. The N_i models are replaced by the single model with the highest local correlation coefficient criterion. The direct sequential co-simulation is performed with the local models of corregionalization—i.e. local correlation coefficients. The conditioning secondary model is now a composition of best parts of the N_i simulations of previous iterations. The composition of a 'seed' image to generate new images in the next iteration follows the crossover principle of the genetic algorithm optimization. The estimation of local means (Eq. 4.6) and variances can be performed in the context of direct co-simulation with bi-distributions

The general outline of this iterative geostatistical methodology can be described in the following sequence of steps, summarized in Fig. 4.4:

(1) Simulate with DSS (Sect. 3.4) at once for the entire seismic grid a set of Ns acoustic impedance models, conditioned to the available acoustic impedance well-log data and assuming a spatial continuity pattern as revealed by a variogram model;

(2) Derive a set Ns synthetic seismic volumes by computing the corresponding normal incidence reflection coefficients (RC; Eq. 4.7) from the impedance models, simulated in the previous step;

Fig. 4.4 Schematic representation of the GSI methodology

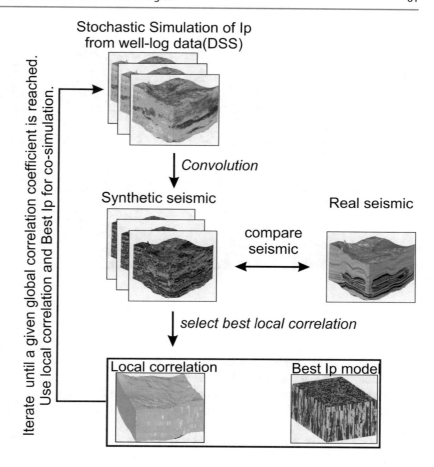

$$RC = \frac{Ip_2 - Ip_1}{Ip_2 + Ip_1}. \qquad (4.7)$$

where the indexes 1 and 2 correspond to the mean above and below the reflection interface considered. The RC are then convolved by an estimated wavelet for that particular seismic dataset in order to compute synthetic seismic volumes (Eq. 1.1).

(3) Convolve these RC with an estimated wavelet for that particular seismic dataset.

(4) Each seismic trace from the Ns synthetic seismic volumes is then compared in terms of correlation coefficient against the real seismic trace from the same location. The comparison is not performed for the entire seismic trace but on a layer basis (Sect. 4.3.6). From the ensemble of simulated Ip models, the acoustic impedance

traces that produce synthetic seismic with the highest correlation coefficient compared with the corresponding real seismic trace are stored in an auxiliary volume along with the value of the correlation coefficient.

(5) These auxiliary volumes, the one with the best acoustic impedance traces and the other with the corresponding local correlation coefficients are used as secondary variables and local regionalized models for the generation of the new set of acoustic impedance models for the next iteration. The new set of Ns acoustic impedance models is built using direct sequential co-simulation (Sect. 3.4.2) conditioned to the available acoustic impedance well-log data, and using the best Ip volumes as secondary variables and local correlation coefficients to condition the joint simulation.

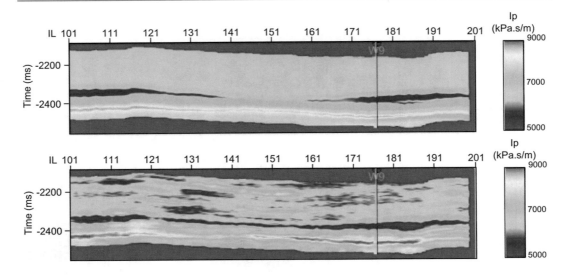

Fig. 4.5 Vertical section extracted from: (*top*) the real Ip model; (*bottom*) best-fit inverse Ip model

(6) The iterative procedure stops when the global correlation coefficient between the full synthetic and real stacked seismic volumes is above a certain threshold (Soares et al. 2007).

The GSI methodology allows the retrieval of high resolution Ip models honoring the distribution function as estimated from the available well-log data, and the spatial continuity model as retrieved from a variogram model. It has been successfully tested on seismic datasets from very different geological contexts with diverse quality.

An application example over a synthetic dataset is shown in Fig. 4.5 (for details of this dataset, please see Sect. 4.3.6). Note that the inverted Ip model is high resolution and matches the main structures of the original Ip model very well.

The use of a global approach ensures that for areas of observed seismic reflection data with poor signal-to-noise ratio the resulting synthetic seismic data remains unmatched throughout the iterative procedure. The lack of convergence results in retrieved Ip models with higher spatial uncertainty for the same locations.

4.3.5 Global Geostatistical Elastic Inversion

The GSI can be extended for the inversion of *n* partial angle stacks directly and simultaneously for acoustic and elastic impedance (Is) models (Nunes et al. 2012). The main purpose of this development was the integration of more information to enrich the final elastic models by simultaneously invert *n* partial angle stacks directly to Ip and Is.

On most occasions Ip and Is do not have a simple linear relationship. Hence, during the model perturbation step of this geostatistical seismic inversion, this inversion algorithm takes advantage of the joint simulation algorithms—co-DSS with joint probability distributions (Sect. 3.4.3)—to reproduce the joint distribution between Ip and Is as retrieved from the available well-log data. The main outline of the GSI (Sect. 4.3.4) is identical to this iterative geostatistical seismic inversion algorithm: the model perturbation is completed by using stochastic sequential simulation, namely DSS and co-DSS with joint probability distributions. At the end of each iteration a genetic algorithm works as a global optimizer of the inversion procedure by maximizing the correlation coefficient between inverted and real partial angle stacks.

The main general outline for the GEI methodology can be briefly described in the following sequence of steps (Fig. 4.6):

(1) Generate of the joint distribution between Ip and Is from the available well-log data;
(2) Simulate a set of *Ns* acoustic impedance models, Ip, with DSS conditioned exclusively to the available acoustic impedance log data and a spatial continuity pattern as revealed by a variogram model;
(3) For each individual Ip model a new elastic impedance model, Is, is generated using co-DSS with joint-probability distributions. The co-simulation of Is, with Ip and best Is models as secondary variable, can be summarized as follows (Fig. 3.5):

(i) The local mean and variance of $Is(x_0)$ is estimated with collocated co-Kriging, from the Is log data and the previously simulated Is values, and with the collocated value of best Is, $Is^*(x_0)$:

$$Is(x_0)^* - m_{Is}(x_0) = \sum_\alpha \lambda_\alpha(x_0)[Is(x_\alpha) - m_{Is}(x_\alpha)] + \lambda_i(x_0)[Is^*(x_0) - m_{Is}(x_0)];$$

(ii) From the bi-distribution (Ip, Is) the conditional distribution of Is, given the collocated value Ip(x_0), is calculated: F (Is|Ip = Ip(x_0));
(iii) A simulated value of Is drawn from the conditional distribution F(Is|Ip = Ip (x_0)), centred in the mean and variance calculated in (i) (Horta and Soares 2010).

At this stage, the Is models are conditioned to the available elastic impedance log data using the previously simulated Ip model as an auxiliary model (Fig. 3.5). Applying this cascade approach of sequential simulation algorithms ensures the

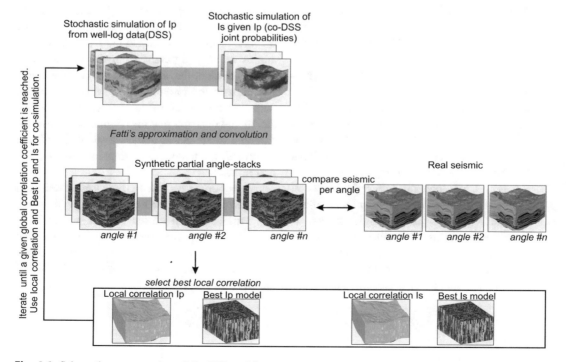

Fig. 4.6 Schematic representation of the GEI workflow

reproduction of the joint-probability distribution between the acoustic and elastic impedance models as estimated from the available well-log data;

(4) From the Ns pairs of Ip and Is, Ns synthetic partial angle stacks are derived using the approximation outlined in Fatti et al. (1994) (Eq. 4.8) for the calculation of the corresponding angle-dependent RC volumes:

$$R_{pp}(\theta) \approx \left(1 + tan^2\theta\right)\frac{\Delta I_p}{\Delta I_s}$$
$$- 4\left(\frac{I_s}{I_p}\right)^2 sin^2\theta\frac{\Delta I_s}{2I_s}, \qquad (4.8)$$

$$\Delta I_p = I_{p2} - I_{p1}$$
$$I_p = \frac{I_{p2} + I_{p1}}{2}$$
$$\Delta I_s = I_{s2} - I_{s1}$$
$$I_s = \frac{I_{s2} + I_{s1}}{2},$$

The index 1 refers to the vertical location in which the calculation of the reflection coefficient is carried out, the mean above the reflection interface; and 2 refers to the sample immediately below, the mean below the reflection interface;

(5) Ns synthetic partial angle stacks are generated by convolution with the corresponding angle-dependent wavelet. It is worth noting that for each individual partial angle stack there must be an estimated wavelet that should be representative of the seismic signal for that specific range of angles;

(6) For each CMP location within the inversion grid, each trace at that particular location, for all the synthetic partial angle stacks, is compared in terms of correlation coefficient with the corresponding recorded seismic trace. From the initial ensemble of duplets of Ip and Is models, the pair of impedance traces that jointly produce a synthetic seismic trace with the highest correlation coefficient when compared with the corresponding observed seismic trace are stored in two auxiliary volumes of best acoustic and elastic impedance models, Ip and Is, along with the respective local correlation coefficients;

(7) The best volumes, together with the local correlation volumes, are then used as secondary variables in the co-simulation process for the generation of a new set of impedance models during the next iteration (Step 3);

(8) The iterative geostatistical inversion procedure finishes when the average global correlation coefficient between the synthetic and the recorded partial angle stacks, for all the available n angle stacks, is above a certain threshold.

This joint simulation assures the reproduction of Ip and Is marginal and joint distributions (Ip, Is) and the spatial continuity patterns, as revealed by the variograms of *Ip* and *Is*, respectively.

The advantage of simultaneously inverting different partial angle stacks is the ability to retrieve coherent elastic and acoustic impedance models. Adding information from reflection angles that are different from the normal incidence enriches the acoustic impedance models. In this way the internal reservoir properties are better described, allowing for better characterization and decision making. Figure 4.7 shows an example of Ip and Is models evaluated with the described methodology, compared against the real models.

When comparing the Ip model retrieved from a simple acoustic inversion (Fig. 4.5) against the one retrieved from elastic inversion (Fig. 4.7), there is clear improvement in the level of detail of the latter. By simultaneously inverting different partial angle stacks the inverse elastic models are richer.

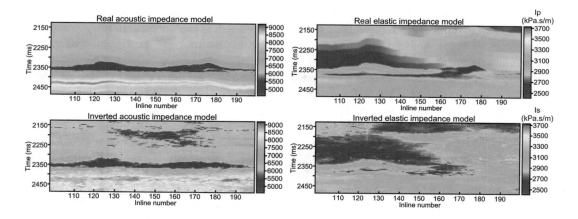

Fig. 4.7 Comparison between a vertical section extracted from a synthetic elastic model and a vertical section extracted from the inverse best-fit model of: (from *left* to *right*) Ip and Is

4.3.6   Geostatistical Seismic AVA Inversion

During the last decade high-quality pre-stack seismic data with high signal-to-noise ratio and considerably high fold has become more available, increasing this data's use in seismic reservoir characterization even within early exploratory stages. The greater availability of high-quality pre-stack seismic data allows for more reliable and less uncertain reservoir models compared to reservoir models derived exclusively from post-stack seismic reflection data. The better subsurface characterization using pre-stack seismic data is achieved by interpreting the changes of amplitude versus the offset (AVO), or with the angle of incidence (AVA; Castagna and Backus 1993; Avseth et al. 2005). The use of pre-stack seismic reflection data is of great importance since it allows the inference of density, P-wave and S-wave velocity models, instead of the traditional acoustic and/or impedance models. Retrieving density and velocity models allows, for example, to better distinguish between litho-fluid facies and to calculate the intrinsic properties of the subsurface geology such as dynamic moduli (Avseth et al. 2005).

Stochastic seismic inversion methodologies for pre-stack seismic data, commonly called seismic AVA inversion, are being proposed based on different assumptions and frameworks (Mallick 1999; Ma 2002; Buland and Omre 2003; Contreras et al. 2005). Here we introduce the geostatistical seismic AVA inversion (Azevedo et al. 2013), which relies on the same general framework of iterative geostatistical seismic inversion algorithms introduced in Sect. 4.3.3.

It is important to note some important points about the seismic processing the pre-stack seismic data requires in order to be successfully inverted (Buland and Omre 2003; Morris et al. 2011): it should, as much as possible, be free of multiple reflections that are not possible to model directly from subsurface elastic models using simple approximations such as Shuey's (1985) linear approximation (Eq. 4.9); the multiple attenuation techniques applied need to preserve, as much as possible, the original relative amplitudes of the recorded seismic data; the CMP gathers should also be corrected for the normal-moveout (NMO) effect with standard seismic velocity analysis. In addition, and after pre-stack migration, the CMP gathers should be corrected for residual NMO effects. Both NMO corrections are critical during the seismic processing sequence, since they ensure the flatness of the gathers to be inverted. Finally, the CMP gathers should be transformed into the angle gathers domain by ray tracing, using a pre-calculated velocity model derived, for example, from traditional velocity analysis. Each trace within an angle gather should refer to a single reflection angle.

As a global geostatistical inversion, the perturbation of the model parameters for density, P-wave and S-wave velocities is performed recurring to DSS (Soares 2001; Horta and Soares 2010).

The convergence of the iterative methodology towards an objective function and the maximization of the global correlation coefficient between real and inverted seismic data, is ensured by returning to a similar global genetic algorithm optimizer based on the cross-over principle. At each iteration, the model parameter space is updated with the elastic portions ensuring the best match between inverted pre-stack synthetic and real seismic data. The best elastic traces can be thought as the best gens at a current iteration and the seed for the generation of new elastic models during the next iteration.

The geostatistical seismic AVA inversion may be summarized in three main stages: stochastic sequential joint simulation of elastic models for

the properties to invert—density, P-wave and S-wave velocities; forward modeling and mismatch evaluation between the observed and the inverted seismic data; and selection of the conditioning data for the generation of the next set of elastic models during the next iteration (Fig. 4.8).

Generating Density, P-wave and S-wave Velocity Models

The stochastic sequential simulation stage comprises the simulation of elastic models following a cascade approach. The sequential procedure starts by simulating, for the entire seismic grid simultaneously, a set of Ns density models followed by the co-simulation of Ns P-wave velocity models from which Ns S-wave velocity models are co-simulated. The order in which each property is simulated and co-simulated is as follows: the density is the first elastic property to be simulated since it is the property associated with a higher degree of uncertainty and its contribution to the recorded seismic reflection data is small. In fact, the component of the seismic

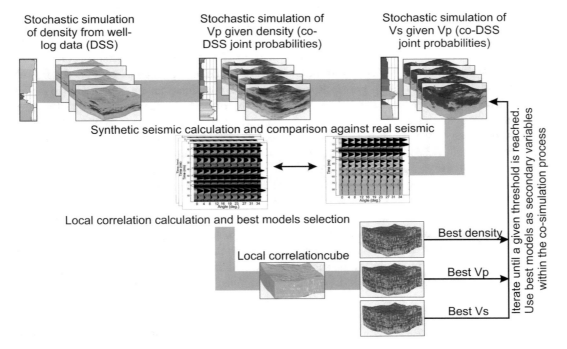

Fig. 4.8 Schematic representation of the iterative geostatistical seismic AVA inversion methodology

reflection data related with density is low and mostly related to the signal received at the far angles (Avseth et al. 2005). Moreover, density is the most spatially homogeneous variable and consequently most convenient to be generated in the first place. Vp and Vs are generated afterwards by stochastic sequential co-simulation.

In the first iteration a set of density models is simulated with DSS conditioned exclusively to the available well-log data, thus ensuring a high degree of variability among the ensemble of simulated models. The level of variability will always depend on the number of available wells against the size of the inversion grid. Each of the Ns simulated density models is then used as conditioning data, along with the available well-log data, for the co-simulation of Ns P-wave velocity models with co-DSS with joint-probability distributions (Sect. 3.4.3). This procedure ensures the reproduction of the bi-distribution between density and P-wave as estimated from the original well-log data.

To conclude the first stage of the iterative inverse procedure, for each of the simulated P-wave velocity models a set of Ns S-wave velocity models is co-simulated, conditioned by the corresponding P-wave velocity model and the available well-log data. The last set of velocity models is simulated using co-DSS with joint probability distributions, ensuring the reproduction of the bi-distribution between both velocities as estimated from the available well-log data.

It is worth noting that at the end of the iterative inversion procedure, the reproduction of the joint distribution densities, Vp and Vs, allows a distinction to be made between any litho-fluid facies previously identified from the original well-log data within the inverted set of elastic models. As well as the spatial interpretation of these litho-fluid facies, the stochastic approach allows the assessment of the spatial uncertainty related with each facies of interest.

Forward Modeling

After the simulation of Ns elastic models, density, Vp and Vs, an ensemble of synthetic pre-stack seismic volumes are calculated.

The angle-dependent RC ($R_{pp}(\theta)$) are calculated following Shuey's (1985) three-terms approximation:

$$R_{pp}(\theta) \approx R(0) + G\sin^2\theta + F\left(\tan^2\theta - \sin^2\theta\right),$$

$$(4.9)$$

with the normal incidence, R(0), reflection as defined by:

$$R(0) = \frac{1}{2}\left(\frac{\Delta Vp}{Vp} + \frac{\Delta\rho}{\rho}\right),$$

and the variation of the reflectivity versus the angle, the AVO gradient, G:

$$G = R(0) - \frac{\Delta V\rho}{V\rho}\left(\frac{1}{2} + \frac{2\Delta Vs^2}{Vs^2}\right) - \frac{4\Delta Vs^2}{Vp^2}\frac{\Delta Vs}{Vs},$$

and F, the reflectivity at the far angles (reflection angles higher than 30°), defined as:

$$F = \frac{1}{2}\frac{\Delta Vp}{Vp}.$$

Each elastic property is defined on each side of the interface where the reflection is happening as follows:

$$\Delta V_p = V_{p2} - V_{p1}$$
$$V_p = \frac{V_{p2} + V_{p1}}{2}$$
$$\Delta V_s = V_{s2} - V_{s1}$$
$$V_s = \frac{V_{s2} + V_{s1}}{2}$$
$$\Delta V_\rho = V_{\rho2} - V_{\rho1}$$
$$V_\rho = \frac{V_{\rho2} + V_{\rho1}}{2},$$

index 1 refers to the vertical location at which the calculation of the reflection coefficient is done, the mean above the reflection interface; while index 2 refers to the sample immediately below, the mean below the reflection interface (Shuey 1985).

By using Shuey's linear approximation within a geostatistical inversion approach we are able to retrieve directly, and as part of the inverse solution, the AVO normal-incidence, *R(0)*, and gradient, *G*, cubes (Rutherford and Williams 1989; Avseth et al. 2005; Castagna and Backus 1993). The propagation of the uncertainty from the inverse problem directly towards the inverted AVO volumes, allows better risk evaluation on amplitude anomalies of interest, which may be related to real hydrocarbon accumulations. However, it is important to note that any alternative approximation to compute the angle-dependent reflection coefficients can be used.

Mismatch Evaluation and Optimization

Each angle trace is composed by *n* seismic traces, equal to the number of reflection angles considered. The *Ns* angle-dependent reflection coefficient traces are convolved by estimated angle-dependent wavelets for each particular incident angle (θ; Fig. 4.9) to obtain *Ns* synthetic angle gathers.

After the forward modeling is complete, the resulting *Ns* synthetic angle gathers are compared with the corresponding real ones in terms of a correlation coefficient. The correlation is performed trace-by-trace per angle gather allowing assessment of local mismatches between synthetic and real traces.

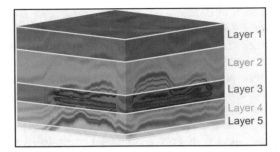

Fig. 4.10 Horizontal layering of the seismic volumes in a random number of horizontal layers with variable vertical sizes. The layering is performed at the beginning of each iteration during the iterative geostatistical seismic AVA methodology

The correlation coefficient between synthetic and observed seismic traces is calculated locally, i.e. the real and the synthetic seismic volumes are divided in a set of horizontal layers (Fig. 4.10). The correlation coefficient between real and synthetic seismic reflection data is then calculated by comparing the synthetic against the real portion of the seismic trace within each layer. This layering approach provides greater local correlation coefficients with a reduced number of simulated elastic models during the first stage of the inverse procedure.

The resulting correlation coefficient is then stored in a new volume composed by local correlation gathers (Fig. 4.11). This procedure is applied to all the layers and to the *Ns* synthetic seismic volumes derived from the ensemble of *Ns* simulated elastic models. The final output of the comparison between observed and synthetic pre-stack seismic data is an ensemble of *Ns* local correlation gathers (Fig. 4.11).

After the mismatch evaluation is complete, the conditioning data used to constrain the co-simulation of the new set of elastic models during the next iteration is then generated (Fig. 4.8). This step comprises the selection and generation of the best density, P-wave and S-wave velocity models along with the corresponding best local correlation cubes. The resulting local correlation coefficients are used by the global optimizer, a genetic algorithm, to converge the inversion into the solution.

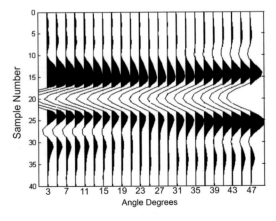

Fig. 4.9 Example of an angle-dependent wavelet, for 23 angles, used for the convolution of the angle-dependent RC ($R_{pp}(\theta)$) during the geostatistical seismic AVA inversion

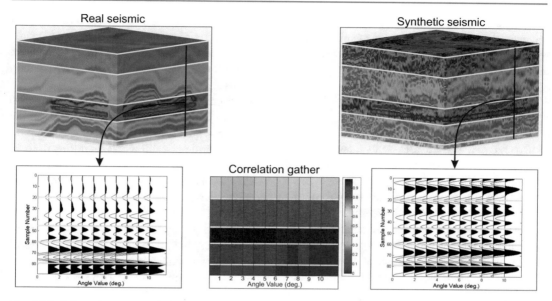

Fig. 4.11 Schematic representation of the comparison procedure between the synthetic and the real pre-stack seismic data. Each portion of trace belonging to each synthetic angle gather, for all the locations within the seismic volumes, is cross-correlated with the corresponding real seismic trace for the same layer. A new correlation volume, composed by correlation gathers, is created with the resulting local correlation coefficients

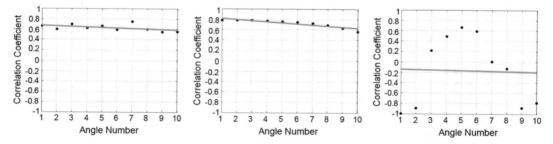

Fig. 4.12 Linear regression fit (*green line*) of the correlation coefficients (*black circles*) for a given layer from three different angles gather realizations at the same spatial location. The elastic models producing the correlation coefficients from the figure on the left are selected as conditioning data for the next generation of models since they ensure the highest correlation coefficient for all the angles simultaneously

Multi-variable Objective Function

The best elastic models, created at the end of each iteration, are composed by the portions of the elastic traces, from the ensemble of density, P-wave and S-wave velocity models, simulated at the current iteration, that jointly produce synthetic seismic reflection data with the highest correlation coefficient compared with the real seismic volume. The selection of this group of elastic traces is not trivial since the correlation coefficient is not the same for all the angles within the same layer (Fig. 4.11). Let us represent the correlation coefficient, r, by reflection angle θ: $r(\theta_i), i = 1, \ldots, N_s$. The selected best elastic traces should be those ensuring the highest correlation for all the angles simultaneously.

If we perform a linear regression for all the values $r(\theta_i)$ (Fig. 4.12), the best simulated elastic traces would be those that ensure the flattest regression (slope of regression equal to zero), the highest intersect value (p) and the minimum least-square deviation value (*LS*). Equation 4.10 give us a global measure *r* of mismatch for a given layer, involving all N_s local correlation coefficients by angles:

$$best\,elastic\,traces = w_1 slope + w_2 p + w_3 LS, \tag{4.10}$$

where w_1, w_2, w_3 are user-defined weights that can vary depending on the quality of the recorded seismic data. If, for example, the far angle has a lower signal-to-noise ratio, the weight of the intersect value, w_2, can be increased.

The best elastic traces will determine the selection of the corresponding Vp, Vs and density, which gave rise to them at that iteration. This is the principle of genetic optimization algorithms. Hence, for a specified layer the local portions of the density, P-wave and S-wave velocity models that produce the most correlated synthetic pre-stack seismic reflection data for all the *n* angles simultaneously are stored in new volumes denominated by best density, P-wave and S-wave velocity models (Fig. 4.13). At the same time, a weighed correlation coefficient for that specific portion of seismic trace is also stored in three different best correlation coefficient volumes for density, P-wave and S-wave velocity respectively (Fig. 4.14).

Each pair of best elastic and local correlation cubes are then used as secondary variables for the stochastic sequential co-simulation of the corresponding elastic properties created during the next iteration. From the evolution of the best elastic models we are able to assess the evolution of the iterative geostatistical methodology (Fig. 4.15). At the end of first iteration, the resulting best model has a patchy appearance (Fig. 4.15 on the left) since it is built from portions of elastic traces from different realizations comprising the first simulation ensemble. As the iterative procedure converges, the resulting best elastic impedance tends to reproduce a spatial continuity pattern towards the inverse solution (Fig. 4.15).

The iterative procedure is considered complete when the global correlation coefficient between the entire synthetic and real pre-stack seismic volumes is above a pre-defined value.

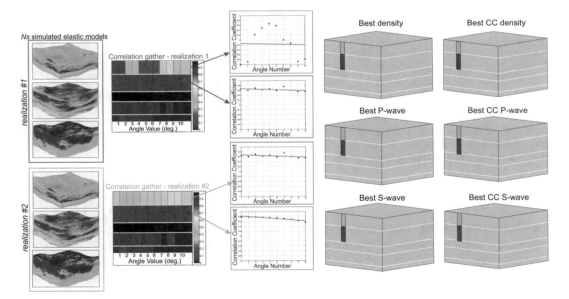

Fig. 4.13 Schematic representation of the procedure to calculate, at the end of each iteration, the best density, P-wave and S-wave velocity models and the corresponding local correlation cubes. These cubes will be used as secondary variables in the co-simulation of the elastic models for the next iteration. Portions of traces selected from realization 1 are represented in *red*, while those selected from realization 2 are plotted in *blue*

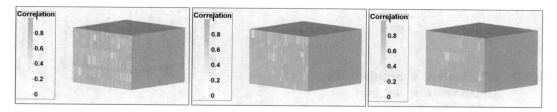

Fig. 4.14 Example of the evolution of the local correlation model. From left: local correlation model at the end of iteration 1; local correlation model at the end of iteration 2 and local correlation model at the end of iteration 6

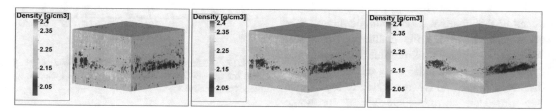

Fig. 4.15 Example of the evolution of the best model for density. From *left*: best model at the end of iteration 1; best model at the end of iteration 2 and best model at the end of iteration 5. Note the patchy effect on the model from iteration 1 is attenuated during the iterative process

In summary, the geostatistical seismic AVA inversion methodology can be described in the following sequence of steps:

(1) Stochastic sequential simulation of *Ns* density models conditioned to the available well-log data with DSS;
(2) Co-simulation of *Ns* P-wave models given the Ns previously simulated density models (co-DSS with joint-distributions);
(3) Stochastic sequential co-simulation of *Ns* S-wave velocity models given the Ns previously simulated P-wave velocity model (co-DSS with joint-distributions);
(4) Calculation of the *Ns* synthetic pre-stack seismic cube with the simulated triplet (density, P-wave, S-wave) using Shuey's linear approximation (Eq. 4.9). In this way the AVO R(0) and G cubes are immediately retrieved as part of the inverse solution;
(5) Compare each synthetic angle gather with the corresponding real gather on a trace-by-trace basis;
(6) Create *Ns* local correlation coeficient gathers for each location within the seismic grid;

(7) Based on a genetic algorithm, select the areas of higher correlation value, from the correlation cube between real and synthetic seismic reflection data, are selected to build the best density, P-wave and S-wave models, which are then used as secondary variables for the co-simulation of the elastic models generated during the next iteration;
(8) Iterate and start (1) until the matching criteria, global correlation between the original and the synthetic pre-stack seismic reflection data, is reached.

Since the model perturbation is based on stochastic sequential simulation, all the elastic models simulated throughout the iterative procedure reproduce the following: the joint probability distributions of density versus P-wave velocity and their marginal probability distributions, the joint probability distributions between P-wave and S-wave velocities and their marginal probability distributions as estimated from the available well-log data, the value of each property at the well locations and the spatial continuity models of density, Vp and Vs as revealed by variogram models.

4.3.7 Application Example with Geostatistical Seismic AVA Inversion

This section describes a small 3D synthetic dataset that has been used to illustrate the different geostatistical inversion procedures introduced in this chapter. This synthetic dataset was constructed to mimic the challenges of a real Atlantic deep-water turbidite field. The available synthetic dataset includes a set of 32 wells with density, P-wave and S-wave velocity logs; noise free pre-stack seismic data sorted by angle gathers with 10 angles uniformly distributed between 0° and 34° (Fig. 4.16); and 3D density, P-wave and S-wave velocity models for the whole study area. An angle-dependent wavelet, with which the synthetic pre-stack seismic data was created, was also included in the available synthetic dataset (Fig. 4.17). The reservoir grid has a size of 101 × 101 × 90 cells in -*i*, -*j* and -*k* directions, respectively (Fig. 4.18). The vertical dimension of each cell is equal to the sample rate of the seismic data (4 ms). The well-log data was built in order to reproduce complex real reservoir

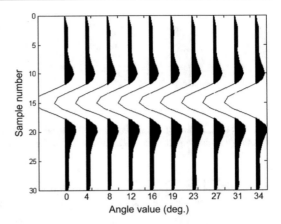

Fig. 4.17 Synthetic angle-dependent wavelet used to create the pre-stack synthetic seismic data and used in the inversion procedure

properties from the analogous turbidite field. From the set of 32 wells only 15 were used as conditioning data for the iterative geostatistical pre-stack seismic inversion. The remaining 17 wells were excluded from the conditioning data and were used exclusively to perform local blind tests to evaluate the performance of the inverse methodology (Fig. 4.18).

From now on, and to better distinguish from the data retrieved by the inversion procedure, the

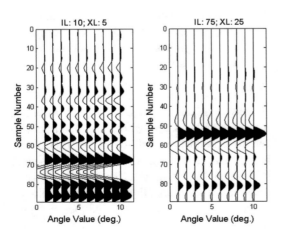

Fig. 4.16 Real seismic angle gathers at two different locations. The seismic signal is considerably variable within the study. In general, the amplitude content increases at the far angles of the reservoir zone

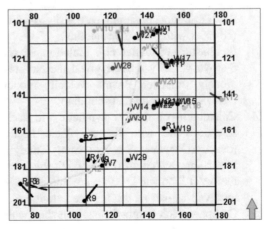

Fig. 4.18 Available set of wells and their location within the seismic grid. Colored wells were used to constrain the geostatistical inversion while *black filled* wells were used exclusively as blind tests. The *yellow line* represents the location of the vertical N–S sections shown to compare the inversion results to the real models

described synthetic data will be designated as real data: real pre-stack seismic data and real subsurface elastic models.

Due to differences in the vertical scale between seismic and well-log data, and in order to keep the synthetic example as realistic as possible, the well-log data was upscaled into the reservoir grid prior to the inversion. The upscaling ensures the reproduction of the main statistics (mean, variance and extreme values) as estimated from the original well-log data (Fig. 4.19).

From the pre-stack seismic data, four partial angle stacks with angles comprehended between 0°–10°, 10°–20°, 20°–30° and 30°–40° respectively were also calculated. The interpretation of the partial angle stacks allows for a better assessment of the spatial complexity and variability of the sedimentary structures present within the study area (Fig. 4.20). In these volumes, the reservoir area is delimited by two strong seismic reflections between 2350 and 2450 ms, corresponding to reservoir's top and base, respectively. The reservoir represents a turbidite channel with an approximate N-S direction. This complexity and heterogeneity can be easily identified in the real density, P-wave and S-wave velocity models (Fig. 4.21). They show important large and small-scale features of interest. For a reliable seismic reservoir characterization, the reproduction of these

variations in the inverted models is extremely important. The reservoir area is characterized by a thin layer of low values of density, associated with low P-wave and S-wave velocity values (Fig. 4.21). It is also important to highlight the presence of local features of interest that should be reproduced in the inverted elastic models, such as a low P-velocity layer around 2000 ms and a high S-wave velocity zone around well R2 (Fig. 4.21).

As in real datasets, the location of the available wells is not randomly distributed along the study area (Fig. 4.18). In fact, most of the wells are located within reservoir or sand-prone areas. The preferential location of the wells is directly translated in a bias of the marginal distributions of the elastic properties of interest (density, P-wave and S-wave velocities) retrieved from the available well-log data when compared with those estimated from the entire real 3D elastic models (Fig. 4.22). The distributions estimated from the set of conditioning wells hardly reproduce the maximum and minimum values or the proportions of each geological facies as estimated from the entire three-dimensional elastic model.

The joint distributions between density versus P-wave velocity and P-wave versus S-wave velocity are complex (Fig. 4.23). It is important to note that these relationships need to be reproduced among inverted elastic models for a

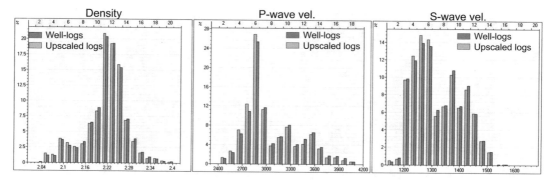

Fig. 4.19 Comparison between the histograms of the original well-log data and the well-log data after the upscaling into the reservoir grid. From *left*: density, P-wave and S-wave velocities. The main statistics (mean and variance) are preserved after the upscaling process

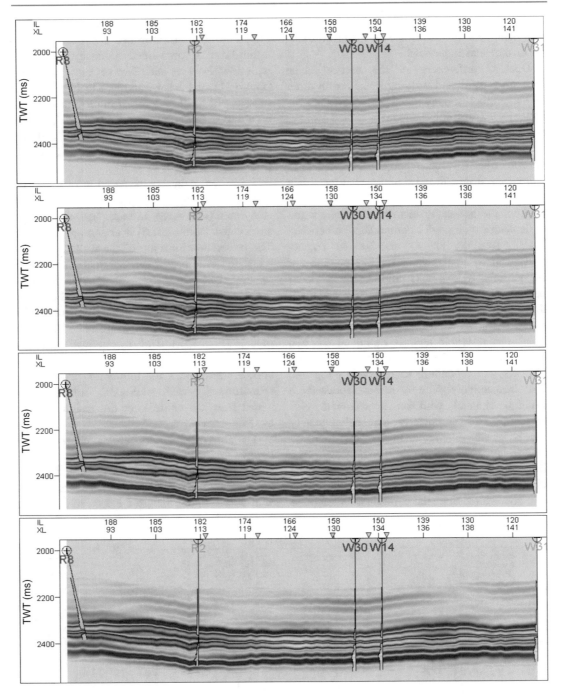

Fig. 4.20 Vertical seismic sections from the real partial angle stacks between: (from *top* to *bottom*) 0°–10°, 10°–20°, 20°–30° and 30°–40°. The location of the vertical seismic profiles is shown in Fig. 4.18. P-wave velocity log is displayed at the well locations. The main seismic reflections between 2350 and 2450 ms are the reservoir's top and base, respectively

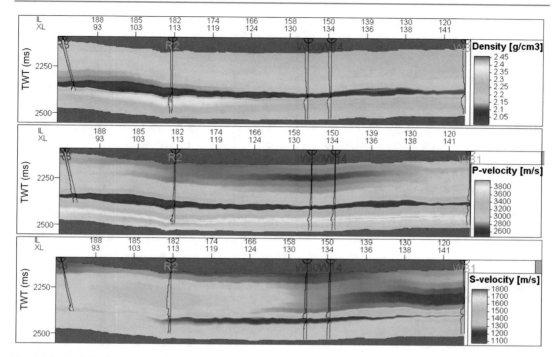

Fig. 4.21 N–S vertical sections from the real elastic models (from *top*: density, P-wave and S-wave velocity models). For location of the vertical section see Fig. 4.18. The reservoir area is defined by low values of density and P-wave velocity, and high values of S-wave velocity

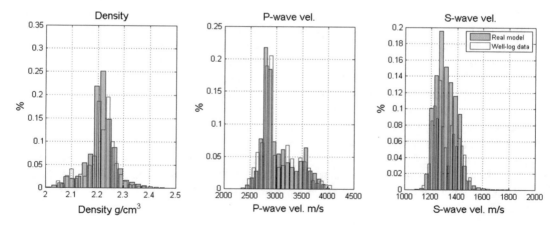

Fig. 4.22 Comparison between the marginal distributions estimated from the real petro-elastic models and the ones estimated from the set of conditioning wells. From *left*: density, P-wave and S-wave velocities. It is clear the well-log data do not perfectly reproduce the original distributions

reliable reservoir characterization. The use of direct sequential co-simulation with joint probability distributions (Horta and Soares 2010) ensures its reproduction between primary and secondary variables.

Spatial Continuity Patterns of Elastic Properties

The spatial continuity pattern of each elastic property was estimated using variogram models estimated for the horizontal and vertical

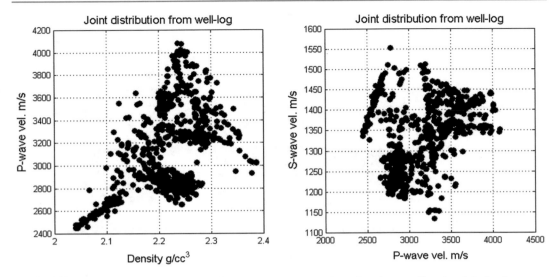

Fig. 4.23 Joint distributions from the set of conditioning well-log data after the upscaling for, *left*, density versus P-wave velocity and *right*, P-wave velocity versus S-wave velocity

directions independently (Fig. 4.24). The experimental variograms were calculated exclusively with the set of well-log data used to constrain the inversion procedure after the upscaling of the original well-log data. Due to the small number of wells and the large distances between well locations, the horizontal experimental variograms are hardly retrieved. For this reason, an omnidirectional horizontal variogram was selected to model the horizontal spatial continuity of each property. On the other hand, the vertical variograms are easy to calculate due to the large number of samples provided by the well-log data. Both directions were modelled with a spherical variogram model with only one structure (Fig. 4.24). These variograms were used for the sequential simulation algorithms, which are part of the geostatistical inversion procedure.

In general, the spatial continuity pattern of all properties is fairly similar. The horizontal variograms do not reach the sill (the variance estimated from the set of experimental data) due to different zonation in the real density, P-wave and S-wave velocity models. These zones correspond to the reservoir itself, its overburden and underburden. The zonation indicates that the elastic

properties we are trying to model are anisotropic and non-stationary in terms of spatial distribution for the entire study area. A zonal anisotropy variogram (Goovaerts 1997) is a suitable approach for conveniently modeling the natural dispersion of these properties. All three elastic properties follow the same spatial continuity pattern. The modelled variograms have horizontal ranges of about 20 grid cells, while the vertical ranges were modelled between 20 and 25 grid cells.

Geostatistical Seismic AVA Inversion: Results

The AVA inversion procedure described in Sect. 4.3.6 was implemented following the simulation sequence of the elastic properties: first, the density was generated by stochastic sequential simulation; P-wave velocity was co-simulated afterwards, conditioned to density; and finally the S-wave velocity was co-simulated conditioned to P-wave pre-simulated values. The inverse procedure converged after six iterations reaching a final global correlation coefficient between the synthetic inverted seismic and the real seismic volumes of 0.75 (Fig. 4.25). All the

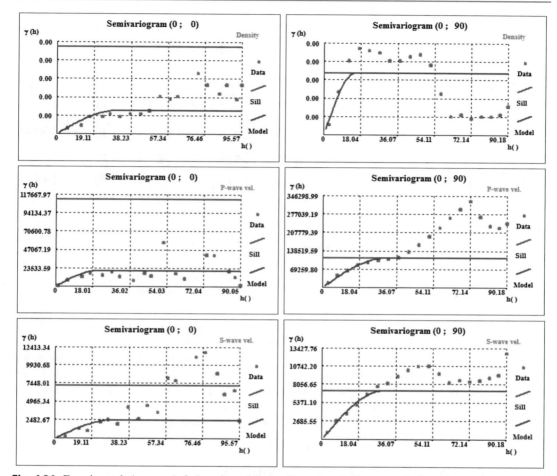

Fig. 4.24 Experimental (*green circles*) and modelled variograms (*blue line*) for the omnidirectional (on the *left*) and vertical directions (on the *right*). From *top*: density, P-wave and S-wave velocity. The variograms were calculated exclusively using the set of conditioning well data

elastic models simulated during the last iteration produced synthetic seismic data with very similar correlation coefficients compared with the real seismic reflection data. On each iteration 32 ensembles of density, P-wave and S-wave velocity models were sequentially simulated recurring to DSS and co-DSS with joint probability distributions. It is common in this kind of geostatistical seismic inversion procedure for the global correlation coefficient between synthetic and real seismic data in the reservoir area to be higher. The main reason for this discrepancy is connected to the way the wavelet is estimated. Normally, the wavelet is estimated to be representative only of the seismic data comprehended

between top and base reservoir. In addition, the convolutional model is only valid after the synthetic RC are convolved by a half wavelet. For this reason, it is best practice to add half the size of the wavelet to the reservoir's top and base surfaces when defining the vertical interval of the inversion grid. In these extra areas the correlation coefficients between real and synthetic seismic reflection data will always be smaller, while the maximum correlation coefficient for the area of interest is ensured.

The convergence of the inverse methodology can also be assessed by the interpretation of local correlation cubes resulting from the trace-by-trace comparison between the real seismic and the

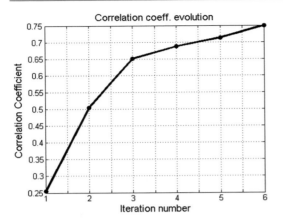

Fig. 4.25 Correlation coefficient evolution at the end of each iteration for the geostatistical seismic AVA inversion example

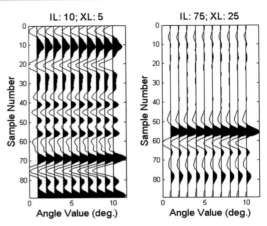

Fig. 4.27 Synthetic seismic angle gathers retrieved at the end of the iterative inversion procedure at the same locations as the real gathers shown in Fig. 4.16. There is a good match between real and synthetic gathers in terms of main reflections and amplitude variation versus offset

synthetic seismic volumes (Fig. 4.26). The interpretation of the local correlation cubes allows the identification of local areas in which the inverted seismic reflection data did not converge and remained with low correlation coefficients. Those areas are often related to low signal-to-noise ratio and consequently should not be matched by the inverted elastic models, or areas where the estimated wavelet is not representative of the observed seismic data.

The best-fit synthetic seismic data is a good match for the real pre-stack data (Fig. 4.27). The synthetic pre-stack seismic data is able to reproduce both the location of the main primary reflections and, more importantly, the amplitude variations versus the offset, or angle. Note that all the models resulting from the last iteration produce synthetic seismic data that is very well correlated with the real one. The resulting mean

model (average of all the models generated during the last iteration) is able to reproduce both the small and large scale details as interpreted from the real elastic models (Fig. 4.28). Particular attention should be paid to the simultaneous reproduction, in terms of values and continuity, of the reservoir and cap rock areas (around 2500 ms) for the three inverted properties.

It is clear that the best retrieved elastic property is the P-wave velocity. The mean P-wave velocity model is able to reproduce the reservoir area as originally interpreted in the real P-wave velocity model fairly continuously. In addition, the cap rock, defined by high values of P-wave velocity, is also very well defined in the resulting P-wave velocity mean model. The worst match between real and inverted mean models is in

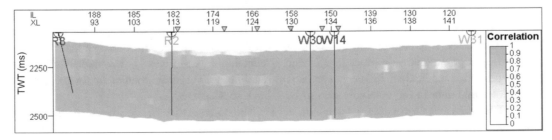

Fig. 4.26 Vertical section extracted from the local correlation *cube* for the synthetic seismic reflection that produce the maximum correlation coefficient when compared with the real seismic data. For location see Fig. 4.18

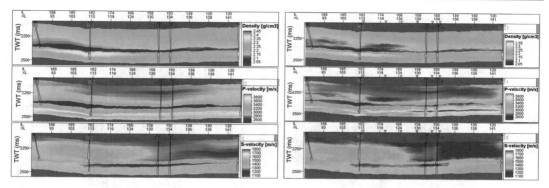

Fig. 4.28 Comparison between vertical sections extracted from (on the *left*) real elastic models and (on the *right*) the mean model computed from the ensemble of models simulated during the last iteration of the geostatistical seismic AVA inversion. From *top*: density, P-wave and S-wave velocity models. The inverted models are constrained by the available well-log data and honor the well data at its location. The inverted mean models are a good match for the real elastic ones. For profile location see Fig. 4.18

those areas less constrained by the well data and above the reservoir's top. The low P-wave velocity model located around 2250 ms near wells W30 and W14 is also present in the inverted models, but more spatially distributed when compared with the real one.

The density mean model is also a considerably good fit for the real density model. The reproduction of the original model is particularly interesting for the reservoir zone and the layers immediately below the reservoir's base. Also, the spatial extent and continuity of the features of interest are present in the inverted mean model.

On the other hand, the mean of the inverted S-wave velocity models simulated during the last iteration succeeds to reproduce the main features present in the real S-wave velocity model. The fit between inverted and real models is reasonable if large scale structures of interest are considered (e.g. the high S-wave velocity area around well R2). However, the small-scale details are hardly interpreted in the inverted S-wave velocity models. The reservoir zone appears more discontinuous, as does the low S-wave velocity layer below the reservoir's base.

It is also important to note that the mean inverted models from Fig. 4.28 reproduce the spatial distribution of the original elastic properties, and are simultaneously able to reproduce its values and their relative variation within the

areas of interest. The small-scale details, which are extremely important for a reliable reservoir characterization, are particularly well retrieved for the reservoir zone as well as for the layers immediately below it.

All the models simulated during the iterative geostatistical inversion procedure for all the elastic properties considered are able to reproduce: the values of the conditioning data at their locations (Fig. 4.28); the joint and marginal distributions (Figs. 4.29 and 4.30) of density, P-wave and S-wave velocities as estimated from the set of conditioning well data; and the spatial continuity models of each property imposed during the sequential simulation by the variogram models (Fig. 4.31).

In order to assess the local convergence of the inverted models, a common procedure is to test these models at specific locations against wells not used as constraining data (i.e. blind wells; Fig. 4.18). Figure 4.32 shows the comparison for wells W19 and W29: the real model (plotted with black solid line) and the mean model computed from all the elastic models simulated during the last iteration (plotted with red dashed line). The inverted elastic models show a fairly good match along the entire well profile. These two wells were selected for blind tests since they are located away from the rest of the conditioning data.

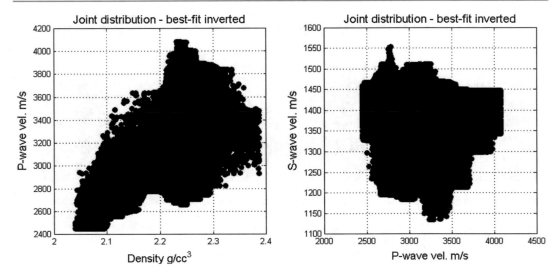

Fig. 4.29 Joint distributions estimated from the best-fit inverted models between: (*left*) density versus P-wave velocity and (*right*) P-wave versus S-wave velocity. They reproduce the joint distributions as estimated exclusively from the well-log data (Fig. 3.5)

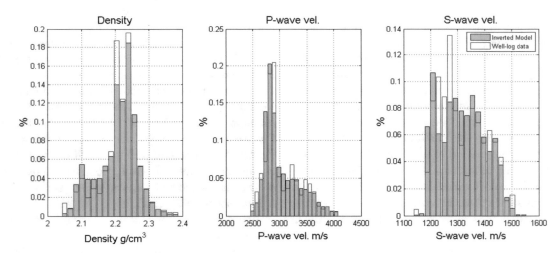

Fig. 4.30 Comparison between the marginal distributions of density, P-wave and S-wave velocity estimated from the conditioning well-log data (*red*) and the ones retrieved from the best-fit inverted elastic models (*green bars*)

Another advantage of using a geostatistical framework to solve the seismic inversion problem is the ability to individually assess the spatial uncertainty of each inverted property. The spatial uncertainty can be assessed by the interpretation of the local correlation cubes resulting from the comparison between synthetic and real seismic traces (Fig. 4.26) or by calculating the variance between the set of inverted elastic models

generated during the last iteration (Fig. 4.33). In the resulting variance model areas of high variability are related with more uncertainty about the model parameters when compared with areas of low variance. Spatial uncertainty can be the result of the reservoir's variability in terms of its internal properties (e.g. geological discontinuities, faults) or the result of a lack of knowledge of the reservoir, such as a lack of or unreliable

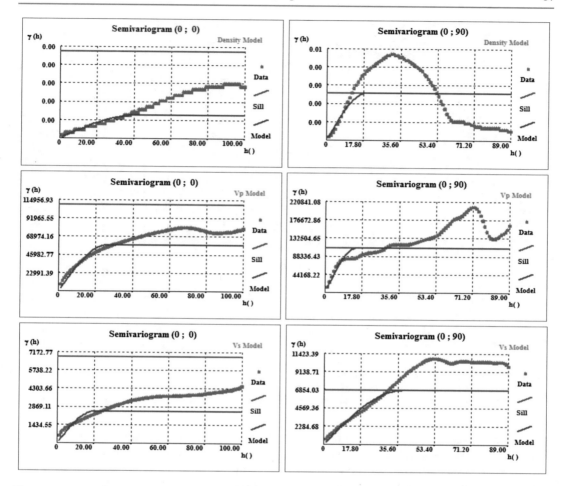

Fig. 4.31 Experimental (*green circles*) and modelled variograms (*blue line*) for the omnidirectional (*left*) and vertical directions (*right*). From *top*: density, P-wave and S-wave velocity. The variograms were calculated over the grids of the best-fit inverted models (compare with Fig. 3.5)

data (e.g. seismic reflection data with a low signal-to-noise ratio). The lack of data is usually approached by the generation of several scenarios for the model parameters and spatial uncertainty is assessed from the sensibility analysis of those scenarios. In these zones there are multiple elastic models that produce synthetic seismic data that fit the real seismic reflection data. High variance values are often related to portions of the observed seismic data with a lower signal-to-noise ratio. The proposed methodology does not force a match with the noise content of the recorded seismic reflection. Even at the end of the inversion procedure, these areas will have elastic models with considerable variability,

producing synthetic seismic data that is unable to converge with the real data.

The variance calculated between the ensemble of elastic models simulated during the last iteration for the case study in the previous section is shown in Fig. 4.33. The P-wave velocity is the property showing lower variance, or lower spatial uncertainty, while the S-wave velocity is the inverted property associated with more uncertainty. The greater variability among the inverted models is preferably located in the southern part of the models since it has less constraining data.

As a seismic AVA inversion methodology, we are also interested in retrieving the AVO normal incidence, R(0), and AVO gradient, G,

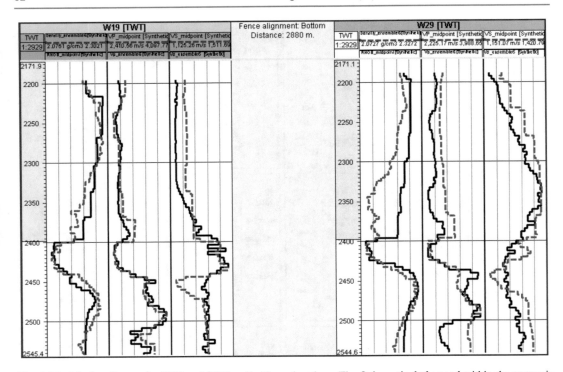

Fig. 4.32 Blind well tests for W19 and W29 wells (for location see Fig. 4.18). The inverted models (*red dashed line*) match the real ones (*black solid line*) at these locations. The fit is particularly good within the reservoir area between 2400 and 2450 ms. From *left*: density, P-wave velocity and S-wave velocity

volumes as defined by Shuey (1985) (Eq. 4.9). For the proposed methodology, these AVO attributes are directly derived for each triplet of simulated density, P-wave and S-wave velocity models and are an intrinsic part of the inverse solution.

The interpretation of the AVO anomalies is traditionally performed in R(0) versus G cross-plots (Fig. 4.34). In this domain AVO anomalies can be classified in types and interpreted in terms of their geological meaning (Rutherford and Williams 1989; Castagna and Backus 1993; Avseth et al. 2005). The classification of AVO responses in this domain was first introduced for gas sands by Rutherford and Williams (1989), who proposed a classification in three classes. This was developed by Castagna and Swan (1997) into four classes. Class I AVO anomalies are those values located in the fourth quadrant of the domain defined by R(0) and G.

They are often related with events with high impedance and low Vp/Vs ratio when compared with the cap rock. Class II are events frequently associated with sands and dim spots in the original seismic reflection data. They are plotted in the fourth and third quadrant. Class III AVO anomalies are classical anomalies related to soft sands filled with hydrocarbon, which are associated with bright spots and plot in the third quadrant. Finally, class IV AVO are rare and are associated with soft sands filled with gas capped with stiff shales. They are plotted in the second quadrant of the R(0) versus G domain (Avseth et al. 2005).

In order to assess the performance of the geostatistical seismic AVO inversion methodology for seismic AVO analysis, an AVO classification cube was calculated from the real three-dimensional elastic models. This cube was compared with the AVO classification cube

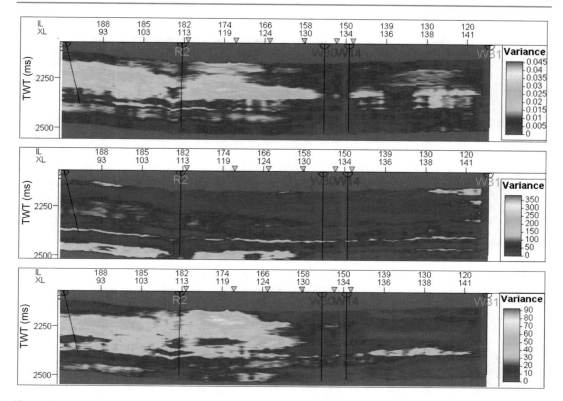

Fig. 4.33 Vertical section of the variance volume computed from the ensemble of simulated models during the last iteration from: (from *top*) density, P-wave and S-wave velocity models. For location of the section see Fig. 4.18

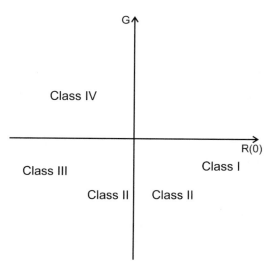

Fig. 4.34 AVO normal incidence, R(0), versus AVO gradient, G, cross-plot and AVO anomalies classification

created from the mean elastic models computed from the ensemble of inverted models simulated during the last iteration of the geostatistical seismic AVO inversion (Fig. 4.35). The inverted AVO cube is able to reproduce the spatial location of the main AVO anomalies; however, the class IV AVO seems to be under-represented in the inverted AVO classification cube. Note that this cube is also part of the inverse solution and conditioned by both the well and seismic reflection data.

An additional feature is the possibility of assessing the uncertainty related with a given AVO anomaly for a specific spatial location. The inverse solution allows the simultaneous plotting of the AVO responses from a set of elastic models in the same R(0) v G cross-plot. This allows distinguishing between anomalies of

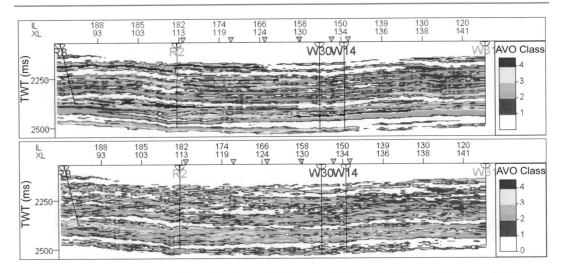

Fig. 4.35 Vertical section extracted from the: (*top*) real AVO classification cube; (*bottom*) AVO classification cube resulting from the mean elastic models simulated during the last iteration. For location of the profile see Fig. 4.18

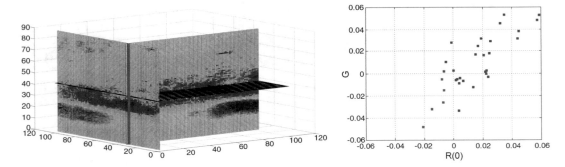

Fig. 4.36 Random vertical and horizontal sections extracted from the best-fit inverted density models and vertical location where the AVO interpretation is being performed (*red line*). Cross-plot between R(0) and G values at the location represented by the intersection of the *red line* with the horizontal density section for the entire ensemble of simulated models during the last iteration. It can be seen that even with inverse elastic models converged towards the reality there are many possible AVO responses

interest related to the presence of hydrocarbons from those anomalies that may be related exclusively to geologic effects while the uncertainty of the AVO response can be assessed (Fig. 4.36).

Deriving Dynamic Moduli

Finally, there is the potential for inversion methodologies based on pre-stack seismic reflection data deriving reliable dynamic moduli such as bulk modulus (K), shear modulus (G), compressional modulus (M) and Poisson's ratio (σ) (Mavko et al. 2003). Deriving reliable dynamic moduli parameters is of extreme importance, for example, for the characterization of the horizontal and vertical stress indexes of a given study area, or for pore fluid discrimination. The comparison between real dynamic moduli (derived from the real elastic models) and the inverted ones (derived from the mean model of the ensemble of models simulated during the last

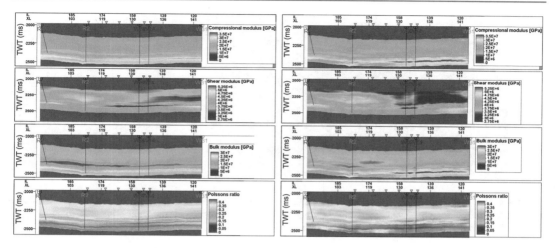

Fig. 4.37 Dynamic moduli calculated from the real elastic models (*left*) and from the mean elastic model computed from the inverted models during the last iteration (*right*). From top: compressional modulus; shear modulus; bulk modulus; and Poisson's ratio

iteration) shows that the inverted models are able to match both large and small scale variations (Fig. 4.37).

4.3.8 Seismic Inversion with Structural Local Models

All geostatistical inversion methodologies introduced above share common characteristics that are intrinsic to the model perturbation technique (Chap. 2): the stationarity assumption about the spatial continuity model, as described by a variogram model, and the probability distribution functions of the properties to be estimated over the entire study area. However, these assumptions can hardly be verified for large study areas and for complex and channeled geological settings in which the sedimentary environments vary rapidly both laterally and vertically: i.e. non-stationary geological environments.

In contexts in which the stationarity assumptions are hardly valid for the entire inversion grid, this can be overcome by using stochastic sequential simulation and co-simulation with local multi-distributions and local spatial continuity patterns. The entire study area should be divided into sub-regions in which the assumption of stationarity is more likely to be valid.

This family of sequential stochastic simulation algorithms reproduces the marginal and joint probability distributions both from the entire set of experimental data and from the data within a given zone inside the inversion grid.

The direct sequential algorithm with multi-distribution and continuity pattern follows the same sequence of steps as introduced in Sect. 3.4. The local mean and variance of a point, located at x_0, is estimated with the local models of variograms. The simulated value is drawn from a local distribution estimated from the experimental data located within that zone. It is worth noting that the spatial continuity pattern is not conditioned by the limits of a particular zone. In fact, when calculating the Kriging estimate and variance at location x_0 the zonation has no influence when searching for the neighbour data points. Finally, the simulated value is drawn from the local distribution of the cell grid to be simulated.

The definition of the spatial zones may be derived directly from seismic interpretation, by well-log interpretation in depth or by both simultaneously (Fig. 4.38).

The vertical zonation illustrated by the example in Fig. 4.38 is easily visually recognized in seismic reflection sections, and in terms of the means and variances from the original S-wave velocity log (Fig. 4.39).

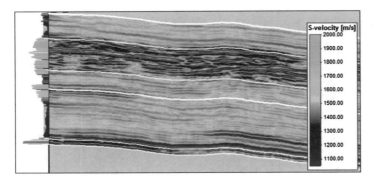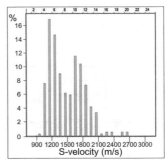

Fig. 4.38 Real case example in which five vertical zones, delimited by the white lines on the *left*, are easily identified in both the vertical seismic section and the S-wave velocity log. The global distribution of S-wave velocity is plotted on the *right*

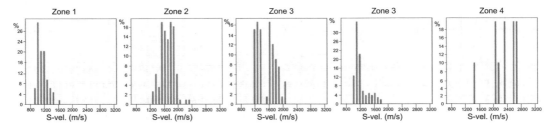

Fig. 4.39 Marginal local histograms for S-wave retrieved from the division shown in Fig. 4.38

In order to compare the impact of including local multi-distributions and variograms within the inversion procedure, we turn to the inverted models obtained by applying of the conventional GSI (Sect. 4.4.2) considering three distinct vertical zones and including DSS with multi-distributions and variogram models within the inversion procedure (Fig. 4.40).

By spatially conditioning the inverse models in terms of distributions and variogram models, we are able to improve the best-fit inverse model since by constraining the spatial location of expected values for the elastic property of interest. In the example from Fig. 4.40 we only expect low Ip values (related with the reservoir zone) in the central part of the model, while high Ip values are mainly constrained by the zone below the reservoir. Including this a priori information allows the retrieval of more reliable Ip models with better matches in terms of the spatial distribution of the subsurface properties of interest.

We also show a real case application of this methodology using geostatistical seismic AVA

inversion. The mean model of the elastic properties co-simulated during the last iteration of the geostatistical inversion is shown in Fig. 4.41. By inverting seismic reflection data with distinct zones (Fig. 4.38), the values of the inferred elastic properties will not appear in areas in which they are not expected (Fig. 4.41). Note that these results are not constrained by any kind of low frequency a priori model. However, by integrating these zones the inversion results are better constrained by previous geological knowledge provided, for example, by seismic and geological interpretation (e.g. it is usual to expect a given litho-fluid facies at a maximum depth).

4.4 Integration of Low-Frequency Models into Geostatistical Seismic Inverse Methodologies

It is widely accepted that, because they allow the integration of data with very different scale support while assessing the spatial uncertainty of

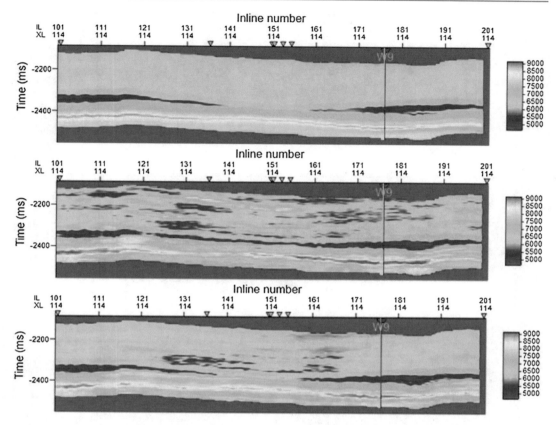

Fig. 4.40 Comparison between vertical sections extracted from the: (*top*) real Ip model; (*middle*) Ip model inverted with the traditional GSI; (*bottom*) Ip model inverted with local multi-distribution functions and variogram models

the inverted properties, geostatistical inversion methodologies may bring more value compared to deterministic inverse approaches. However, geostatistical inversion methodologies are very computationally expensive compared to deterministic approaches.

Also, there is a frequent discussion within the geophysical community that for most of the geostatistical seismic inversion methodologies, as those discussed above, there is no real control of the initial low-frequency model (LFM) that fills the frequency spectrum below the seismic resolution (frequently between 5 and 10 Hz; Fig. 4.1). The initial LFM used to constrain an inversion procedure has a crucial impact on the retrieved inverse elastic model. In fact, most of the deterministic inverse solutions based on LFM

search the model parameter space around the initial model with various kinds of optimization technique, e.g. conjugate gradients, simulated annealing. If this initial guess is far from the global minimum, then the inversion procedure will struggle to find local minima close to the global solution.

The common approach to model an LFM is by Kriging the well-log data of the property of interest for the bandwidth of interest and constraining this interpolation by any available interpreted horizon. The interpolation follows the topography of key surface, such as the top and/or base reservoir. In order to constrain the interpolation, it is usual to have a processing velocity model as a trend, using Kriging with trend, for example (Dubrule 2003). The use of LFM allows

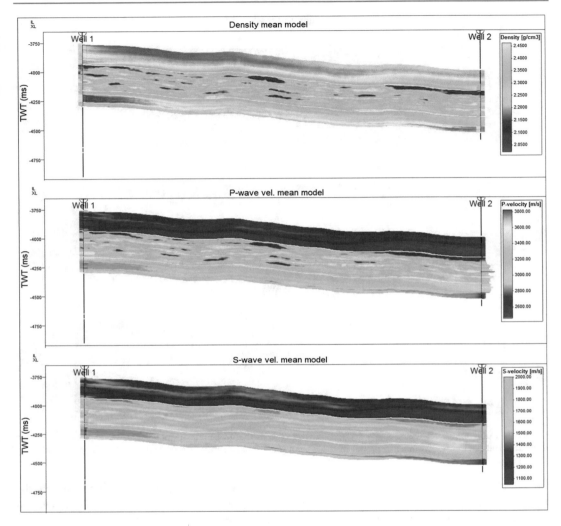

Fig. 4.41 Vertical section through well locations from the mean model of density, P-wave and S-wave velocities resulting from the geostatistical seismic AVA inversion with multi-local distributions and spatial continuity patterns

faster and cheaper solutions, since using an initial LFM considerably constrains the exploration of the model parameter space.

This section proposes a methodology to explicitly incorporate a LFM, or a priori initial guess model, into geostatistical seismic methodologies. The integration of this model is part of the objective function used in the iterative procedure. The traditional objective function based exclusively on the correlation coefficients between real and synthetic seismic traces is combined with the deviations between

the initial guess model (LFM) and each single model generated during a given iteration, individually for each elastic property of interest individually. The combined objective function (Eq. 4.11) drives the global optimizer used as part of iterative geostatistical seismic inversion techniques ensuring the convergence of the iterative procedure from iteration to iteration. This approach has a direct impact on the match between real and synthetic seismic data and convergence rate of geostatistical seismic inversion methodologies:

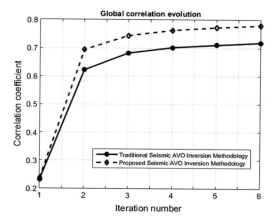

Fig. 4.42 Global correlation evolution comparison between traditional and proposed seismic AVA inversion methodologies. It shows that the proposed technique has a faster and higher convergence

$$OF = \lambda_1[\rho(synthetic, real)] + \lambda_2[m - m^*],$$
$$(4.11)$$

where $\rho(synthetic, real)$ is the trace-by-trace correlation coefficient between synthetic and real seismic, m is the LFM for the property of interest and m^* a single realization for the same property. λ_1 and λ_2 are user-defined weights that associated with each term of the objective function.

It is worth noting that by adding an a priori LFM to the objective function we are restricting the uncertainty space of the model parameter space and increasing the convergence rate. This compromise can be tuned through the weights, λ_1 and λ_2, of the objective function.

The application of this integrated approach on a real case study using geostatistical seismic AVA inversion directly for facies (Azevedo et al. 2015) demonstrates that by integrating a priori knowledge this inversion technique has a faster and higher convergence value compared to the traditional geostatistical seismic inversion approach (Fig. 4.42). When comparing the best-fit inverse models resulting from both methodologies, this integrated procedure results, in this case, in better spatial continuity and geometry definition of the channels as interpreted in both models (Fig. 4.43). Note that in addition to the better spatial constraint, the inverted elastic models remain high resolution. The high-frequency content is kept by using stochastic sequential simulation and by directly integrating the well-log data within the inversion procedure: i.e. each single model generated during the inversion loop reproduces the well-log data values at the well locations.

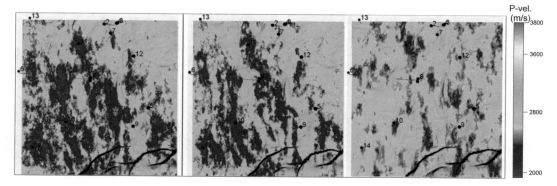

Fig. 4.43 Comparison between traditional seismic AVA inversion (*left*), proposed seismic AVA inversion (*middle*) and the initial guess model (*right*) used in the proposed methodology. Horizontal time section through best-fit P-wave velocity model

Deriving Petrophysical Properties with Seismic Inversion

5

5.1 Characterization of Petrophysical Properties Based on Acoustic and Velocity Models

The geostatistical framework presented in Chap. 2 allows the integration of data with different scale support. From the resulting inverted acoustic and/or elastic models, the use of geostatistical techniques, such as sequential co-simulation (Chap. 3), allows the simulation and co-simulation of the petrophysical properties of interest. The geostatistical simulated models (described above) can use the well-log data of the property to be modelled as the primary variable and the inverted elastic models as secondary variable for the co-simulation (or co-estimation) of the petrophysical property of interest. In this sequential approach, petrophysical modeling is performed in two independent steps. The petrophysical property of interest is inferred from the acoustic or elastic model obtained from the inversion procedure; therefore, the resulting petrophysical model is not directly constrained by the available seismic reflection data.

Co-simulation of petrophysical properties based on acoustic impedance models

The traditional workflow for inferring the subsurface petrophysical properties of interest is based on a joint simulation (Chap. 2) with an elastic model previously retrieved from the

observed seismic data by seismic inversion as a secondary variable. The degree of detail (i.e. small-scale variability) of the resulting petrophysical model depends on the inverse methodology selected among the many available methods.

In this traditional sequential workflow, a relationship between the elastic and the petrophysical property of interest is first derived from the available well-log data. Depending on the geological setting of the study area, the relationship between these two properties may be linear or extremely complex, representing different geological formations, pore types and pore fluids (Fig. 5.1), for example.

The petrophysical model is then derived by resorting to stochastic sequential co-simulation using the available well-log data for the property of interest as experimental data and the collocated inverted elastic model as secondary variable. The secondary variable is often the best-fit inverted model or the mean model calculated from the set of models generated during the last iteration of an iterative geostatistical seismic inversion procedure. Depending on the joint probability distribution between both properties, we may use the traditional sequential co-simulation with a global correlation coefficient between both properties, for linear relationships between the primary and the secondary variable (Sect. 3.4.2), or the sequential co-simulation with joint probability distributions for complex relationships (Sect. 3.4.3). Nevertheless, in either case the resulting

© Springer International Publishing AG 2017
L. Azevedo and A. Soares, *Geostatistical Methods for Reservoir Geophysics*,
Advances in Oil and Gas Exploration & Production, DOI 10.1007/978-3-319-53201-1_5

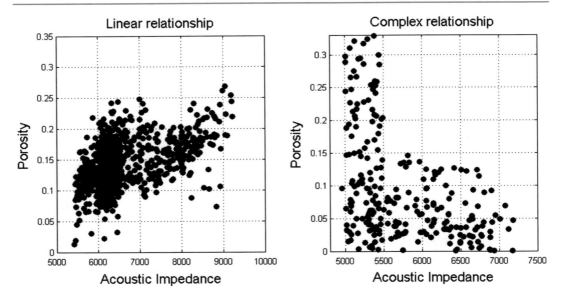

Fig. 5.1 From *left* example of an approximately linear relationship between Ip and porosity and a complex relationship between Ip and porosity

petrophysical model is not directly retrieved from the seismic reflection data and consequently is not directly constrained by this data.

When compared with petrophysical models derived exclusively from well-log data, the uncertainty related with these models smaller; however, they are not directly constrained by the seismic reflection data and, locally, they may disagree with the observed seismic data. New inversion methodologies that permit directly inferring petrophysical models of interest from the available seismic reflection data allow for more realistic and well-constrained subsurface Earth models: i.e. subsurface models with less uncertainty.

5.2 Direct Inversion of Porosity Models

In seismic reservoir characterization, the classical procedure for obtaining a petrophysical model of interest, such as a porosity model, is sequentially inverting seismic reflection data for an elastic property (e.g. acoustic and/or elastic impedance) and then inferring the petrophysical property of interest. The conversion between the elastic and

the petrophysical domains is based on a calibrated relationship model between acoustic and/or elastic impedance and the petrophysical properties of interest (e.g. porosity; e.g. Avseth et al. 2005). In these techniques, seismic reflection data is integrated through a forward model, frequently a stochastic sequential simulation algorithm, for example, sequential Gaussian co-simulation (Deutsch and Journel 1992) or DSS (Soares 2001). The stochastic sequential simulation can account for the differences in support, but cannot accommodate a nonlinear transformation of original variables, as in the sequential Gaussian co-simulation (Gómez-Hernandez and Journel 1993).

In this approach we do not directly constrain the resulting petrophysical property model to the available seismic reflection data: porosity is normally derived linearly from a specific seismic attribute, the acoustic and/or elastic impedance model, retrieved by seismic inversion (e.g. Doyen 2007). Consequently, this kind of approach is normally plagued by several problems related to: first, the different (space/time) support of the seismic data versus the measurements of the internal properties of the reservoir at the well locations; and, second, in complex geological environments the fact the nonlinear relationship between the two variables, elastic and petrophysical properties, are the product of the joint behaviour of several distinct geological facies, makes it difficult to retrieve with such a sequential approach and its extrapolation is unlikely to be valid for the entire study area.

In such complex geological environments, different facies of interest cannot be modelled solely on the basis of their acoustic and petrophysical properties (Fig. 5.2) and an assumption of linearity can lead to erroneous and unreliable models. Several alternative approaches have been proposed to tackle this problem, for example, by using artificial neural network algorithms to infer the elastic properties from facies models (Sun et al. 2001) or by the prior characterization of facies data (Robinson 2001). Alternative methods consist of obtaining subsurface elastic models by seismic inversion, after which porosity models are derived through a calibrated RPM between the inverted elastic property and the petrophysical properties of interest (Mavko et al. 2003; Bosch et al. 2010).

Here we introduce a methodology to jointly characterize acoustic impedance and porosity by resorting to direct sequential co-simulation with joint probability distributions (Sect. 3.4.3). This methodology can be extended easily to any other petrophysical property of interest as long as it has

a relationship with acoustic impedance. Using co-DSS with joint probability distributions to derive porosity models within the inversion loop ensures that the complex relationships between Ip and porosity are reproduced, the derived porosity models are directly constrained by the available seismic reflection data and the uncertainty is propagated during the inversion procedure towards the final porosity models.

The integration of porosity models within geostatistical seismic inversion may be summarized as follows:

(1) Generate an initial set of porosity models from the available well-log data;
(2) Co-simulate, using direct sequential co-simulation with joint probability distribution, acoustic impedance models for the whole study area (Horta and Soares 2001) from the available well-log data and using the models generated in (1) as auxiliary variables. The example shown here considers acoustic impedance, but its extension to other geostatistical seismic inversion methodologies is straightforward;
(3) Calculate the corresponding synthetic seismic volumes by convolving the reflectivity series derived from the simulated acoustic impedance models with an estimated wavelet representative of the entire field;

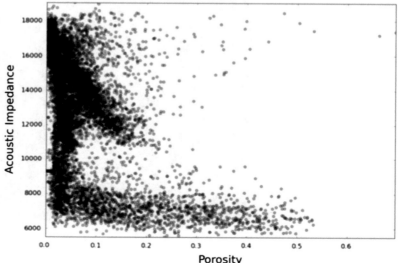

Fig. 5.2 Example of complex relationship between porosity and acoustic impedance where different geological facies can easily be distinguished by the petro-elastic domain defined by acoustic impedance versus porosity

(4) Evaluate the mismatch between the entire set of synthetic seismic data created in the previous step and the recorded seismic data by calculating local correlation coefficients in a trace-by-trace basis;

(5) From the set of acoustic impedance models from a given iteration, select the elastic traces that ensure the highest correlation coefficients between the corresponding synthetic and real seismic traces and store them in an auxiliary volume with the corresponding correlation coefficients and the porosity traces;

(6) Generate a new set of porosity models by direct sequential co-simulation using the auxiliary volumes of porosity and the corresponding correlation coefficient volume created in the previous step. Return to step (1) until the objective function, the global correlation coefficient between real and synthetic seismic data, reaches a given threshold.

These kind of approaches, based on DSS algorithms, ensure the reproduction of marginal and joint distributions as given by the co-variogram for both the elastic and the petrophysical property of interest. With this approach, we retrieve petrophysical model, directly constrained by the available seismic reflection data. This is a crucial improvement for obtaining more reliable reservoir models, since the joint and marginal distributions of both properties, as revealed by the experimental data, are spatially reproduced in the resulting models.

The potential of inverting directly the available seismic reflection data for porosity is shown with acoustic inversion in a case study with data from an early exploration area located on a complex turbiditic environment. The study area was first divided vertically into three different zones (Fig. 5.3). Zone 2 corresponds to the channel system of interest in which the prospect is located.

Fig. 5.3 Joint distributions between Ip and porosity for the three zones defined along the inversion grid

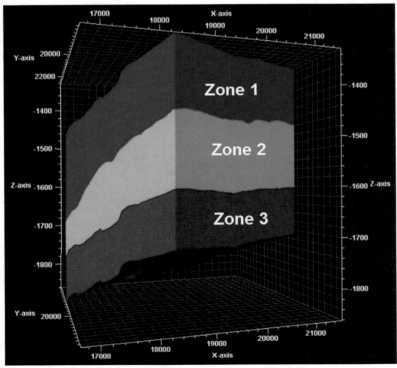

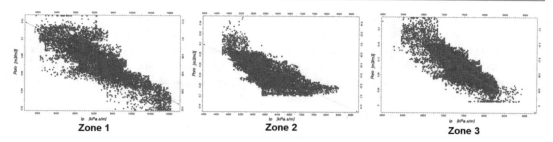

Fig. 5.4 Vertical regionalization of the study area in which the global geostatistical inversion integrating porosity models was implemented

A local joint distribution between Ip and porosity, based on well-log data available for the study area and nearby fields, was assigned to each zone. While zones 1 and 2 have a linear relationship between these, the properties for zones 2 exhibit different behavior (Fig. 5.4).

The best-fit pair of Ip and porosity retrieved at the end of six iterations (Fig. 5.5) show a consistent spatial distribution of both properties. By directly inverting for porosity by integration seismic reflection and well-log data within this geostatistical seismic inversion framework, we allow the inference of high-resolution models for both properties and ensure the uncertainty related with seismic inversion is simultaneously propagated towards both models.

5.3   Geostatistical Seismic Inversion Directly for Petrophysical Properties

In seismic reservoir characterization studies we normally seek to infer the detailed spatial distribution of the subsurface facies while conditioning the petro-elastic properties to such a model. In this section we introduce geostatistical inversion methodologies that allow both the inversion of seismic reflection data for elastic properties and the simultaneous inference of the spatial distribution for the petrophysical properties of interest as they are revealed in a rock physics model.

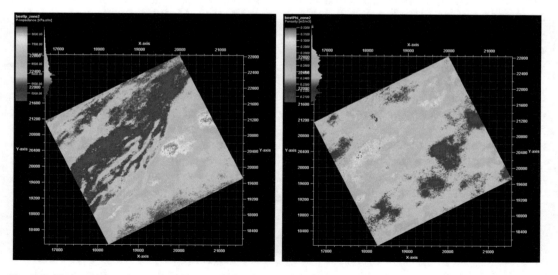

Fig. 5.5 Horizon slices extracted within zone 2 for the best-fit inverse model of: *left* Ip and *right* porosity

The integration of rock physics models within seismic inversion methodologies allows for more realistic reservoir models, bringing the characteristics of the subsurface geology such as mineralogy, pore type, fluids, porosity, sorting, degree of cementation and stress into the inverted petroelastic models. There are different theoretical and empirical rock physics models that can be used to describe the elastic response of a given geological formation as a function its properties. The most typical rock physics models for clastic reservoirs are summarized in Mavko et al. (2003), Avseth et al. (2005), Simm and Bacon (2014) and, for carbonates reservoirs, in Xu and Payne (2009).

5.3.1 Geostatistical Seismic AVA Inversion to Facies

The inversion methodology presented here can be considered an extension of the geostatistical seismic AVA inversion technique (introduced in Sect. 4.3.6) allowing the direct inversion of pre-stack seismic data directly for facies, density, P-wave and S-wave velocity models. The advantage of this methodology is the inference of facies models that are intrinsic to the inverse solution and which are therefore conditioned simultaneously by both: the available well-log data and the pre-stack seismic data. Introducing a facies model within the inversion procedure represents a straighter connection between the elastic properties (observed seismic) and the subsurface geology.

The geostatistical seismic AVA inversion directly to facies is an iterative geostatistical seismic inversion based on three main principles:

(1) The perturbation of the model parameter space is performed sequentially, by stochastic sequential simulation and co-simulation: DSS (Soares 2001) as the perturbation technique for density models and co-DSS with joint probability distributions (Horta and Soares 2010) as the model perturbation procedure for P-wave and S-wave velocity models.

(2) The convergence of the iterative methodology towards an objective function is ensured by a global optimizer based on the crossover genetic algorithm of previous iterations. At each iteration the model parameter space is updated with elastic traces that ensure the greatest correlation coefficient between synthetic and real pre-stack angle gathers at a given iteration.

(3) The use of Bayesian classification (Avseth et al. 2005) to create facies models from simulated and co-simulated pairs elastic models of density and Vp/Vs ratio.

We may summarize this inverse methodology in six main steps (Fig. 5.6). First, the facies of interest are identified in an elastic domain such as the one defined by density versus Vp/Vs ratio domain (Fig. 5.7). This is a step of the utmost importance, since the classification at the well locations will be used as training data for the Bayesian classification that is performed as part of the inversion loop. The success of the inversion procedure is highly dependent on the reliability of this classification and the ability to separate the facies of interest in a given elastic domain.

The second stage concerns the simulation and joint simulation of both variables of the rock physics relation, density and Vp/Vs ratio. First the DSS of *Ns* density models, conditioned to the available well-log data, followed by the direct sequential co-simulation with joint probability distributions of Vp/Vs ratio that describes—along with density—the elastic domain from where the facies were defined in the previous step (e.g. Vp/Vs ratio). Then, from the *Ns* pairs of density and Vp/Vs models, *Ns* facies models are categorized according to a probabilistic classification method, such as the Bayesian classification, and using the pre-calibrated training data at the well locations (Avseth et al. 2005).

Inside each facies, the inverse procedure continues with the stochastic sequential simulation of *Ns* P-wave and S-wave velocity models using DSS and co/DSS with joint probability distributions (as in Sect. 4.3.6). The resulting P-wave velocity models honor both the global probability distribution as estimated from the well-log and the individual probability

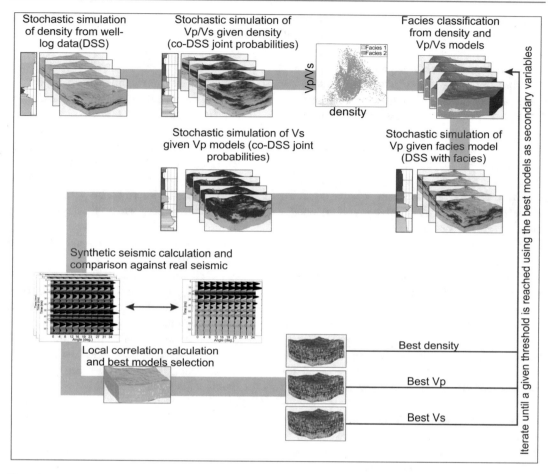

Fig. 5.6 Schematic representation of the geostatistical inversion of seismic AVA data directly to facies models methodology

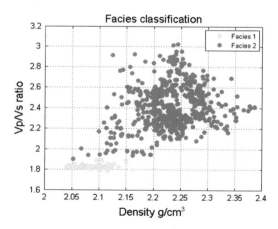

Fig. 5.7 Facies classification, in two different facies, from well-log data in the elastic domain defined by density versus Vp/Vs

distributions for each of the facies of interest as defined by the training dataset.

After the simulation of Ns triplets of density, P-wave and S-wave velocity models, the iterative geostatistical inversion procedure follows the sequence described in the previous section for the geostatistical seismic AVA inversion (Sect. 4.3.6) as in Fig. 5.6.

Angle-dependent RC are calculated, following Shuey's linear approximation (Eq. 4.9), from the set of Ns elastic models. The resulting RC are then convolved with an angle-dependent wavelet, producing a set of Ns synthetic angle gathers. The synthetic and real pre-stack volumes are horizontally layered (Fig. 4.11). Each layer of the synthetic angle gathers is then cross-correlated on a trace-by-trace basis, with the corresponding layer of the real angle gather. The resulting local correlation coefficients are stored in local correlation angle gathers (Fig. 4.13).

From the entire set of elastic models simulated at the current iteration, the triplets of density, P-wave and S-wave velocity models, which produce synthetic seismic reflection data with the greatest local correlation coefficient simultaneously for all the angles when compared with the real seismic data, are stored in the best density, P-wave and S-wave velocity models (Fig. 4.13). A local correlation coefficient is also assigned to each portion of the elastic models selected for the best volumes. The correlation coefficients are weighted averages of the local correlation gathers assigned to that particular triplet of elastic models. The correlation coefficients corresponding to the near angles are averaged for the best correlation cube associated to the best P-wave velocity model. The correlation coefficients corresponding to the mid and far angles are averaged for the best correlation cube associated to the best S-wave velocity model. The correlation coefficients corresponding to the far angles are averaged for the best correlation cube associated to the best density model. Note that the best density will be the conditioning data for a new set of density models, the best P-wave for the corresponding P-wave velocity models and, finally, the best S-wave velocity as secondary variables for the co-simulation of a new set of S-wave velocity models. The variable r remains unconditioned for the entire inversion loop.

The best elastic models, and the local correlation volumes, are then used as secondary variables for the sequential co-simulation of elastic models of the next iteration. The iterative geostatistical seismic inversion finishes when the global correlation between synthetic and real seismic is above a certain threshold.

All the inverted models reproduce the main spatial patterns, as revealed by the variograms imposed during the stochastic sequential simulation of the elastic models, the probability distributions and joint probability distributions estimated from the original well-log data and the

well-log data at the well location. The described algorithm may be summarized as shown in Fig. 5.6:

(1) Define facies from available well-log data in the elastic domain defined by: for example, density versus Vp/Vs ratio;

(2) Stochastic sequential simulation of N_s density models conditioned from available well-log data with DSS (Soares 2001);

(3) Stochastic sequential co-simulation of N_s models for the intermediate variable r (e.g. Vp/Vs ratio) using the previously simulated N_s density models as secondary variables (co-DSS with joint-distributions) (Horta and Soares 2010);

(4) For each pair of models generated in steps (1) and (2) classify probabilistic facies (e.g. Bayesian classification) resulting in N_s facies volumes;

(5) Stochastic sequential simulation of N_s P-wave velocity models with the DSS with multi-local distributions using a facies model as auxiliary variable;

(6) Stochastic sequential co-simulation of N_s models of S-wave velocity models given the N_s previously simulated P-wave velocity models (co-DSS with joint-distributions) (Horta and Soares 2010);

(7) For each of the N_s elastic models previously simulated, calculate the pre-stack synthetic seismic data following Shuey's linear approximation;

(8) Compare each synthetic angle gathers with the corresponding real gather on a trace-by-trace basis;

(9) Store the elastic traces of density, P-wave and S-wave velocities that, for a given iteration, produce synthetic angle gathers with the highest correlation coefficient between real and synthetic seismic traces. These elastic traces, and the corresponding correlation coefficient value, are used as secondary volumes for the co-simulation of density, P-wave and S-wave models for the next iteration. Note that the generation of Vp/Vs ratio models during the entire inversion procedure is only conditioned to a

previously generated density model. From iteration to iteration, the conditioning based on the match between real and synthetic seismic data is done on the elastic properties used to compute the synthetic seismic: i.e. the Vp/Vs ratio model and the facies model resulting from the Bayesian classification are not directly constrained by the previous iteration.

(10) Iterate until the matching criteria, the global correlation between the original and the synthetic seismic, is reached.

5.3.2   Application Examples with Geostatistical Seismic AVA Inversion Directly to Facies

We illustrate the potential of the inversion directly to facies with two simple examples: a synthetic and a real application. The first synthetic example uses the same dataset used to illustrate the geostatistical seismic AVA inversion (Sect. 4.3.6). In this way we can directly compare the improvements obtained by integrating facies within the inversion loop. The second example shows a real implementation in a challenging mature turbidite field where previous geostatistical seismic inversion studies are available.

Synthetic example

A facies model was calibrated by using to the 15 available conditioning wells in the elastic domain defined by density versus Vp/Vs ratio. This elastic domain was chosen for its potential in separating facies filled with different fluids (Fig. 5.7): Facies 1 corresponds to the reservoir, while facies 2 is related with non-reservoir lithologies (i.e. the overburden and the underburden lithologies). Facies 1 is related with low values of the Vp/Vs ratio and the low density associated with sands filled with hydrocarbons. Please note that this synthetic example illustrates,

and is solely intended to demonstrate, the potential of this inverse procedure.

This inverse methodology comprises the co-simulation of a new elastic property (besides P-wave and S-wave velocities) from a previously simulated density model. In this example, and due to its potential for separating the litho-fluid facies of interest, the variable r to be co-simulated from the density models is the Vp/Vs ratio. The simulated pair of density and Vp/Vs ratio models is then classified into a facies model by Bayesian classification.

The real Vp/Vs ratio model shows a low velocity ratio values around the 2400 ms related to the reservoir zone. Other potential regions of interest that should be reproduced in the inverted models are related to local velocity variations (e.g. low-velocity layers) such as the one around well R2 before the 2250 ms (Fig. 5.8).

The reference density and Vp/Vs simulated ratio models were classified in a true facies model (Fig. 5.9) following the facies classification performed from the well-log data (Fig. 5.7). The facies corresponding to the reservoir (facies 1) has large lateral variations in terms of its thickness and spatial distribution. Reliable inverted facies models should reproduce these variations or, at the very least, reproduce areas of greater uncertainty for the areas in which its thickness changes dramatically.

The iterative procedure converged after six iterations on each set of 32 elastic models (density, P-wave and S-wave velocity and Vp/Vs ratio) were reproduced. At each iteration, from the 32 pairs of simulated models of density and Vp/Vs ratio, 32 facies models where classified using Bayesian classification. The final global correlation coefficient between the synthetic seismic derived from the best-fit inverted models and the real pre-stack seismic is 0.81 (Fig. 5.10).

The best-fit synthetic seismic data is able to reproduce the main primary reflections and AVA variations as interpreted from the real pre-stack seismic data (Fig. 5.11). From a seismic AVA perspective, it is a key point for ensuring a match between real and synthetic seismic data both in

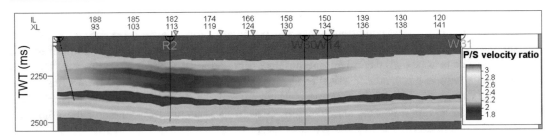

Fig. 5.8   Vertical section extracted from the real Vp/Vs ratio model. For location of the profile see Fig. 4.18

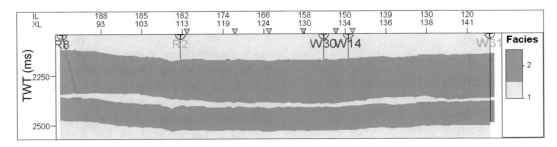

Fig. 5.9   Vertical section extracted from the real facies volume classified from the real elastic models. Facies 1 corresponds to the reservoir area while facies 2 to the non-reservoir. For location of the profile see Fig. 4.18

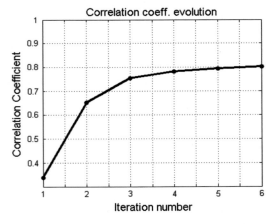

Fig. 5.10   Correlation coefficient evolution at the end of each iteration for the geostatistical seismic AVA inversion directly to facies example

the shape of seismic traces and, more importantly, in the relative variation of the amplitude content from angle to angle.

The local correlation coefficient volume, calculated on a trace-by-trace basis between the synthetic seismic data and from the best-fit inverted elastic models and real seismic data,

shows a good convergence of the inverse procedure (Fig. 5.12).

The convergence of the inverted elastic models can be assessed by comparing the mean model of all elastic models simulated during the last iteration against the real elastic models (Fig. 5.13). All elastic models simulated during the last iteration produce synthetic seismic reflection data with a correlation coefficient of around 0.8 compared to the real seismic data. Both density and P-wave velocity models are particularly well retrieved. In both inverted models there is a very good match for the reservoir and cap rock zones in terms of their spatial continuity and values. However, for the P-wave velocity model, the areas above the cap rock, before 2250 ms, have lower values compared to the real model. It seems the lower P-wave velocity layer, around 2200 ms, has a greater extent than the real P-wave velocity model.

The S-wave velocity model is the elastic model in which the inversion methodology performs worst. Nevertheless, the inverted

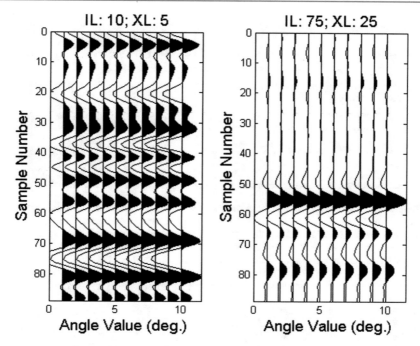

Fig. 5.11 Synthetic seismic angle gathers retrieved at the end of the iterative inversion procedure at the same locations as the real gathers shown in Fig. 4.18. There is a good match between real and synthetic gathers in terms of main reflections and amplitude variation versus offset

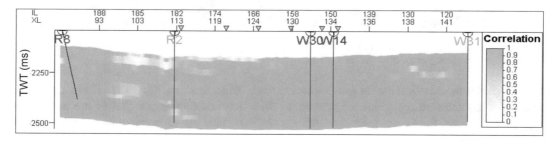

Fig. 5.12 Vertical section extracted from the local correlation cube of the best-fit inverted seismic. For location see Fig. 4.18

S-wave models are able to reproduce the main large-scale features at its locations. Moreover, the inverted S-wave velocity models can reproduce some small-scale events, such as the thin low S-wave velocity layer immediately below the reservoir's base (around 2400 ms). Comparing with the geostatistical seismic AVA inversion (Sect. 4.3.6), there is a better match between the real and the inverted S-wave velocity models. This result is due to the integration of the facies model calibration within the inversion procedure.

The facies models corresponding to the mode of the set of facies models classified during the last iteration of the inversion procedure can reproduce the spatial distribution of both facies as interpreted from the original facies volume (Fig. 5.14). As for facies 1, which corresponds to the reservoir area, the reproduction is good both in its spatial distribution and in terms of reservoir thickness. However, the facies present in the mean facies model are more discontinuous compared to the real facies model (Fig. 5.9). This

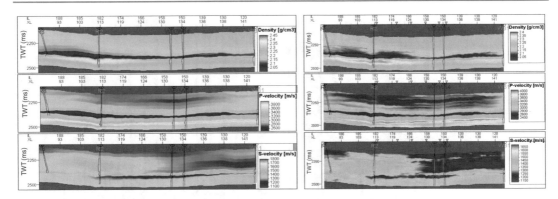

Fig. 5.13 Comparison between vertical sections extracted from (*on the left*) real elastic models and (*on the right*) the mean model calculated from the set of simulated models during the last iteration of the geostatistical seismic AVA inversion directly to facies. From *top*: density, P-wave and S-wave velocity models. The inverted models are constrained by the available well-log data and honor the well data at its location. They can reproduce the main features of the real elastic models. For profile location see Fig. 4.18

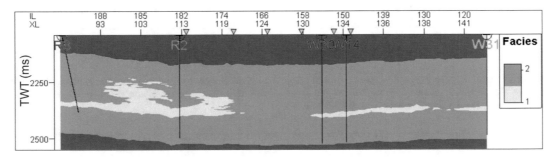

Fig. 5.14 Vertical section extracted from the mode of facies model calculated from the set of facies models simulated during the last iteration. For profile location see Fig. 4.18

discontinuity is more noticeable for locations less constrained by the well data. Nevertheless, it is important to note there are individual facies models, classified during the last iteration, with more continuous facies with better matches to the real facies model (Fig. 5.15). These discrepancies between models are part of the uncertainty related with the reservoir and should be interpreted as an advantage of the proposed methodology.

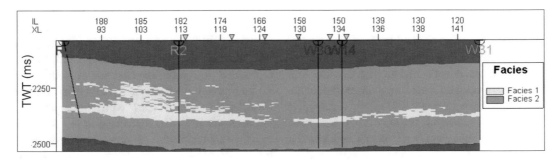

Fig. 5.15 Vertical section extracted from the best-fit facies model. The facies corresponding to reservoir (facies 1) is much more continuous than that resulting from the mean model of the simulated facies models during the last iteration (Fig. 5.14). For profile location see Fig. 4.18

The inverted elastic models are also able to reproduce the joint distributions, as estimated from the well-log, between density and Vp/Vs ratio and Vp versus Vs (Fig. 5.16) while honoring the facies classification as performed in the training data. This is an essential feature of the proposed iterative geostatistical methodology, since it both ensures the reproduction of the relationships between elastic properties and constrains them to a geological model that is calibrated at the well locations (the facies model itself).

Finally, the marginal distributions of the elastic models simulated during the entire inversion workflow (density, Vp/Vs ratio, P-wave and S-wave velocities) reproduce those distribution as estimated from the well-log data (Fig. 5.17).

It is important to note the differences within the retrieved inverse elastic models with and without a facies model (compare Figs. 4.24 and 5.13). It is clear that including a geological link within the inversion procedure considerably improves the inverse elastic models in both the values of the retrieved properties and the spatial interest of the features of interest (Fig. 5.13).

Real example

To illustrate the potential of this method in real datasets, we show the implementation of this geostatistical seismic inversion procedure in a real dataset.

The study area is a deep offshore turbidite environment in which the reservoirs are associated with sand-prone overbank deposits. The known reservoirs are recognized on partial angle stack due to their amplitude anomalies with the offset. While the proposed method was developed to handle angle gathers, because of the lack of pre-stack data we used partial angle stacks with the following central angles of reflection: 10°, 15°, 20° and 29°.

A set of 13 wells with Vp, Vs and bulk density logs were available. The wells drill preferentially the pay geological formations, which introduce a bias on the known elastic properties (Fig. 5.18). These wells were previously tied to the available seismic data and the resulting angle-dependent wavelets were also made available. The high resolution well-log data was upscaled into the reservoir grid, ensuring the extreme values, the mean and the variance as

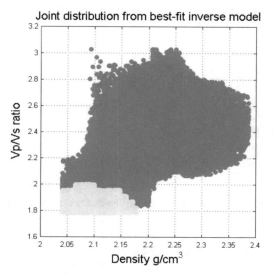

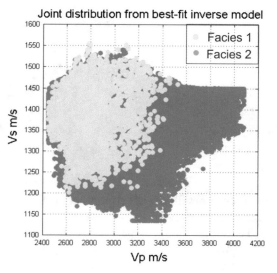

Fig. 5.16 Joint distributions estimated from the best-fit inverted models between: *left* density versus Vp/Vs ratio and *right* P-wave versus S-wave velocity. They reproduce the joint distributions as estimated from the well-log data (Figs. 4.19 and 4.31). The joint distributions are color-coded by facies. The reproduction of the relationships between elastic properties while keeping the geologic realism as provided by the facies model calibrated at the well locations is essential

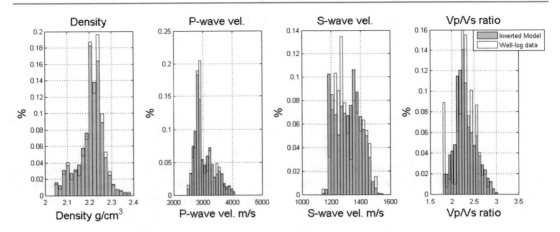

Fig. 5.17 Comparison of the marginal distributions of density, P-wave and S-wave velocity and Vp/Vs ratio estimated from the conditioning well-log data (*red*) and that retrieved from the best-fit inverted elastic models (*green*)

Fig. 5.18 Available set of wells and their locations within the seismic grid for the real case study. The *black dashed line* represents the location of the vertical well sections shown to present the results of the geostatistical seismic AVA for facies. Well head locations are represented by *black circles* and the deviation path by the *thin black line*

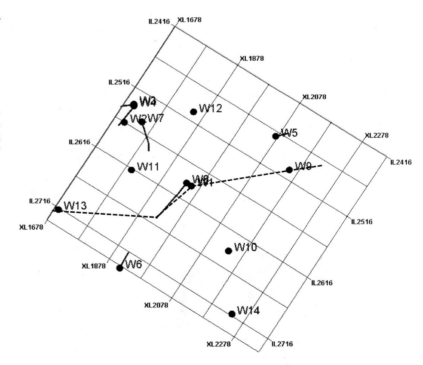

retrieved from the original well-log data was preserved after upscaling.

The spatial continuity pattern of each property was inferred by modeling experimental variograms calculated in the vertical direction from the upscaled well-log data and in the horizontal direction from the real partial angle stacks. Because of the distances between wells we decided to model the horizontal spatial continuity

pattern from the seismic reflection data. This procedure is normally translated in an overestimation of the variogram range values compared with variograms retrieved from well-log data.

We defined two facies of interest prior to the geostatistical inversion from the elastic domain defined by the density and Vp/Vs ratio (Fig. 5.19). This classification enables a distinction to be made between reservoir and

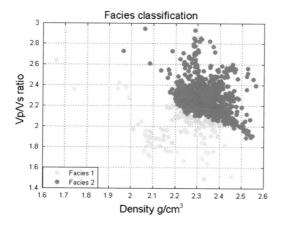

Fig. 5.19 Joint distribution between density and Vp/Vs ratio from the upscaled well-log data and colored by facies type. This was used as training data for the Bayesian classification within the iterative geostatistical workflow. Facies 1 is the reservoir facies while facies 2 is the non-reservoir facies

non-reservoir facies: i.e. sand-prone and shaly lithologies, respectively. This was used as training data for the Bayesian classification included as part of the geostatistical inversion procedure.

The geostatistical seismic AVA inversion converged towards the real seismic reflection data after six iterations. Per iteration, 32 sets of density, P-wave and S-wave velocity were simulated and co-simulated. After six iterations, the global correlation coefficient between the real and synthetic partially stacked seismic reflection for all angles simultaneously is about 0.75.

The retrieved synthetic seismic data has a considerably good match with the real seismic data in both primary reflection and AVA variations. It is important to note that both the amplitude values and the extension of the seismic events of interest as interpreted from the real models are reproduced on the synthetic seismic reflection data (Fig. 5.20).

Due to the similarity of the petro-elastic models generated during the last iteration, all these models produced synthetic seismic data that was highly correlated with the recorded seismic data. The mean model of density, P-wave and S-wave velocity models calculated from the set of elastic models generated during the last iteration can be used to interpret the inversion

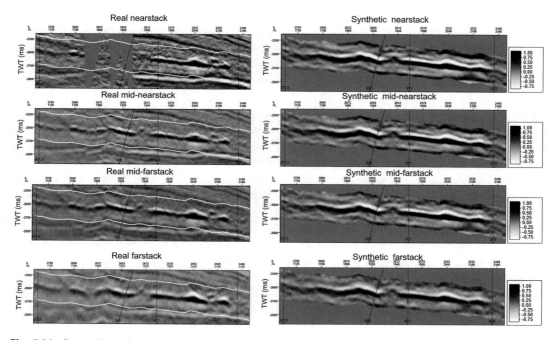

Fig. 5.20 Comparison of vertical seismic sections extracted from: *left* real partial stacks and *right* synthetic seismic reflection retrieved from the geostatistical seismic AVA inversion for facies

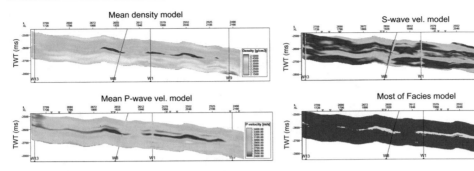

Fig. 5.21 Mean petro-elastic models calculated from the set of models generated during the last iteration of the geostatistical seismic AVA inversion for facies. The

inverted models are constrained by the available well-log data and honor the well data at its location

results (Fig. 5.21). The inverted models are high resolution and show both large and small details of interest. Each model is constrained by the corresponding well-log log data at its locations.

The modal facies model (Fig. 5.21), calculated from the set of facies models generated during the last iteration of the inversion, show the proportion of reservoir facies in accordance with what is known about this reservoir. It is interesting to observe that the reservoir facies is not connected along the whole reservoir area, but that its locations generally agree with the amplitude anomalies interpreted from the real seismic reflection data.

Unlike the geostatistical seismic AVA methodology (Sect. 4.3.6), the inversion of seismic data directly to facies allows retrieval of both the best-fit inverse elastic models and a geological model of the reservoir. Instead of being derived from an inverted elastic model, the resulting inverse facies model is constrained simultaneously by both well-log and seismic data. The integration of the facies models during the inversion process brings more geological realism to the inverse solution. However, this approach can be applied to more complex geological environments by fully integrating other rock physic models (Asveth et al. 2005) into the geostatistical seismic inversion procedure.

5.4 Integration of Rock Physics Models into Geostatistical Seismic Inversion

New developments for iterative geostatistical seismic inversion procedures should integrate statistical rock physics within the inversion workflow. The approach proposed here may be summarized as follows: the first step comprises the calibration of a petro-elastic model from the well-log data. A theoretical or empirical model is fitted by linear or nonlinear regressions to the available well-log data. At the end of this step we are able to derive a set of equations governing the geological system being examined. As in the conventional statistical rock physics workflow, when considering pore fluids not sampled by the available well data, fluid substitution by Gassmann's equation (e.g. Smith et al. 2003) can be performed and included in the calibrated petro-elastic model. The second step adds the variability of the subsurface geology by generating values of the petro-elastic properties of interest not sampled by the well-log data by using Monte Carlo simulation. Adding this variability to the calibrated petro-elastic model is of great importance since it allows more variable stochastic simulated models created during the

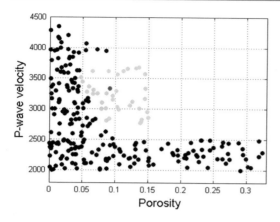

Fig. 5.22 Example of a bi-distribution between porosity and P-wave velocity where several populations, corresponding to different facies, can be distinguished. The *red circle* represents the value for P-wave velocity computed from a petro-elastic model given the collocated value for porosity in x_0. The local bi-distribution between porosity versus P-wave velocity (*blue circles*) will be used in the co-simulation of the P-wave and S-wave velocity models

iterative geostatistical inversion procedure and, therefore, more reliable solutions.

The third stage is the iterative geostatistical inversion algorithm itself as described in Sect. 4.3. The inversion procedure begins with the stochastic sequential simulation of Ns water saturation (Sw) models by DSS from the well-log data. Then, in a sequential approach, using the co-DSS with joint probability distributions, generate Ns models of porosity conditioned to the available well-log data and the previously simulated Sw models. The resulting porosity models are then used as secondary variables for the co-simulation of Ns density models with co-DSS with joint-probability distributions.

At each location within the reservoir grid, x_0, P-wave and S-wave values are drawn from the probabilistic petro-elastic model constructed in step one, taking the collocated simulated values of Sw (x_0) and porosity (x_0) into account. These values will be used to determine the local joint probability distributions between porosity versus P-wave velocity, and porosity versus S-wave velocity (Fig. 5.22). The selected local probability distributions will be used in the direct sequential co-simulation with joint probability distributions for the simulation of Ns P-wave and S-wave velocity models.

The iterative geostatistical inversion algorithm then follows the approach presented for the geostatistical seismic AVA methodology (Sect. 4.3.6).

In this chapter we generalize the geostatistical
seismic inverse methodologies to simultaneously
integrate data of different nature, besides the
conventional seismic reflection and well-log data
as shown in the previous chapter. Here we illustrate recent advances in integrating controlled-
source electromagnetic (CSEM; Sect. 6.1) and
dynamic production data (Sect. 6.2) within the
inversion procedure.

6.1 Integration of Electromagnetic and Seismic Data into Geostatistical Simultaneous Inversion

Recent advances in geophysical methods have
allowed the use of electromagnetic data, particularly controlled-source electromagnetic
(CSEM), to infer the spatial distribution of the
subsurface pore fluid distribution (e.g. Gao et al.
2010; Hoversten et al. 2006). The resistivity
model retrieved from CSEM inversion are later
used to derive petrophysical quantities of interest, such as porosity and water saturation by
using Archie's Law (Eq. 6.1; Archie 1942)
and/or Waxman and Smits' equations (Eq. 6.2;
Waxman and Smits 1968). These laws relate the
electrical conductivity, or specific resistance, of a
rock to its porosity and brine saturation:

$$R = aS_w^{-n}\phi^{-m}R_w,\qquad(6.1)$$

where, R is the specific resistance of the partially
saturated rock at the brine saturation Sw, Sw is

the saturation of brine, and n is the saturation
exponent. ϕ is the rock porosity and m is the
cementation exponent:

$$\sigma_0 = \frac{1}{F}(\sigma_w + BQ_v),\qquad(6.2)$$

where, σ_0 is the electrical conductivity, F is the
formation factor, σ_w is the electrical conductivity
for brine, B is the equivalent conductance of
sodium clay exchange cations, and Qv is the
cation exchange capacity per unit pore volume.

Given Eqs. 6.1 and 6.2, we can look at the
CSEM data as the solution of an inverse problem
with the porosity and brine saturations as
unknown parameters.

CSEM inversion shares the same characteristics of any other geophysical inverse problem.
They are ill-posed problems, nonlinear and with
non-unique solutions due to the limited bandwidth
and resolution of the geophysical data, noise,
measurement errors and physical assumptions
about the involved forward models (Tarantola
2005). The model parameters are updated until the
match between synthetic and real resistivity measurements is achieved. The idea of the joint use of
seismic reflection and CSEM data is to generate
facies, or porosity, models that simultaneously
match both data. By integrating different kinds of
data within the same inversion framework we
expect to be able to reduce this uncertainty level
and consequently retrieve more reliable subsurface Earth models. It is important to note that
when compared with seismic reflection data,
CSEM has considerably lower resolution with

L. Azevedo and A. Soares, *Geostatistical Methods for Reservoir Geophysics*,
Advances in Oil and Gas Exploration & Production, DOI 10.1007/978-3-319-53201-1_6

much lower subsurface penetration (Tompkins et al. 2011). The proposed framework, ensures both the propagation of the uncertainty during the entire inversion procedure and the simultaneous integration of data with very different scale support: well-log, seismic reflection and CSEM data (Bosch et al. 2010).

EM inversion is increasing its importance in inferring subsurface resistivity distribution that is later used to derive the spatial distribution of subsurface fluids (Gao et al. 2010; Hoversten et al. 2006). These methodologies are traditionally applied individually and, due to the non-unique nature of geophysical inversion, the resulting models may not be consistent when jointly interpreted.

By jointly inverting seismic reflection and CSEM data, the retrieved petro-elastic models allow for better reservoir characterization both in terms of the spatial distribution of the petro-elastic properties of interest and by separating areas filled with different fluids. A key aspect of this joint inversion is that both kinds of geophysical measurements have different spatial sensitivity and, consequently, there is no combined influence from the two types of measurements.

In practice, because of the very different nature of these geophysical methods, their simultaneous inversion is not straightforward and, therefore, there is a need to use a rock physics model to link both domains: the elastic domain derived from the seismic reflection data is linked with the petrophysical property derived from the inverted resistivity models.

As an example for assessing the potential of joint inversion in reservoir characterization, here we show an iterative geostatistical simultaneous inversion of CSEM and seismic reflection data. The geostatistical joint inversion of seismic and electromagnetic data allows the simultaneously inference of subsurface acoustic impedance, water saturation and porosity models from available well-log data, seismic reflection and CSEM data (Azevedo and Soares 2014). It is an iterative geostatistical methodology in which the model perturbation is performed by DSS and co-simulation (Sects. 3.4.1 and 3.4.3). Water saturation and porosity data is obtained by

simulating water saturation and porosity by using the previously simulated saturation models as auxiliary variables (DSS with joint probability distributions, Sect. 3.4.3). Each pair is used to simultaneously produce synthetic resistivity data (e.g. Archie's Law) and synthetic seismic data.

The available well-log data act as experimental data for the stochastic sequential simulation and co-simulation algorithms, and a genetic algorithm that is based on the cross-over principle works as a global optimizer that simultaneously converges, with each iteration, the synthetic resistivity and seismic data into the real data. The inversion procedure is considered complete if the average global correlation coefficient between the entire volume of real data and the inverted synthetic resistivity and seismic reflection data, simultaneously, are above a certain user-defined threshold.

The geostatistical joint inversion of seismic and electromagnetic data can be summarized through the following sequence of steps (Fig. 6.1):

(1) Simulate *Ns* models of water saturation (Sw) by stochastic sequential simulation, DSS (Soares 2001) and using the available Sw-log data as experimental data for the simulation procedure;

(2) Co-simulate *Ns* porosity models using DSS with joint probability distributions (Horta and Soares 2010), the available porosity well-log data as experimental data and each Sw model simulated in the previous step as a secondary variable;

(3) Classify a facies model (e.g. sand/shale) for each pair of models created in (1) and (2);

(4) Following Archie's Law (Archie 1942) and the Waxman and Smits equations (1968), calculate *Ns* synthetic resistivity responses for each pair of Sw and porosity models simulated and co-simulated in the previous steps depending on the facies model classified in (3);

(5) Following a pre-calibrated RPM, and for each porosity model generated in (2), derive *Ns* acoustic impedance models and compute the corresponding normal-incidence RC. These RC are then convolved with an estimated wavelet;

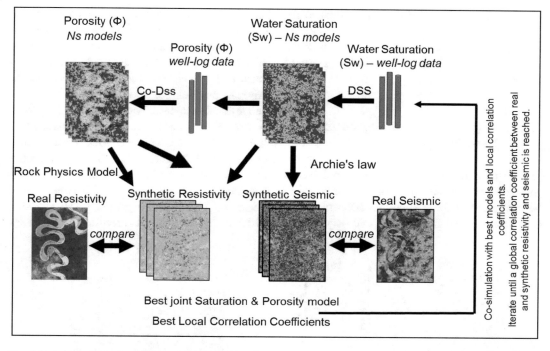

Fig. 6.1 Schematic representation of the geostatistical joint inversion of seismic and electromagnetic data workflow

(6) Calculate, on a trace-by-trace basis, the correlation coefficient between synthetic and real resistivity and seismic responses;

(7) Select the petro-elastic traces that simultaneously ensure the maximum correlation coefficient between recorded and synthetic seismic and resistivity data calculated in the previous step. The individual correlation coefficients are weighted averages, depending on the quality of the input geophysical data. Save these elastic traces as the best joint saturation and porosity volumes along with the corresponding correlation coefficients;

(8) Co-simulate a new set of Sw and porosity models by co-DSS and using the best joint saturation and porosity models as secondary variables along with the available well-log data; Iterate and return to (2) until a given global correlation coefficient between synthetic and real resistivity and seismic data is reached.

As a joint inversion methodology, the convergence of the iterative procedure is ensured to simultaneously match both EM and seismic reflection data through a weighted mean of the individual trace-by-trace correlation coefficient between real and synthetic seismic reflection and resistivity data (Eq. 6.3):

$$CC_{trace} = w_1 * CC_{seismic} + w_2 * CC_{EM}, \quad (6.3)$$

where CC_{trace} is the joint trace-by-trace correlation coefficient between real and synthetic seismic reflection and resistivity data, w_1 is the weight associated with the individual correlation coefficient between seismic traces ($CC_{seismic}$) and w_2 is the weight associated with the individual correlation coefficient between EM traces (CC_{EM}). The weights of Eq. 6.3 may be tuned depending on the noise level of each recorded geophysical data, and the influence of each kind of geophysical data within the inversion procedure.

All petrophysical models generated during the iterative workflow honor the well-log data at its locations, reproduce the marginal probability distributions of Sw and porosity as estimated from the available well-log data, reproduce the joint probability distribution between Sw and porosity as estimated by the available well-log data and reproduce a spatial continuity pattern as in the variogram model.

6.1.1   Application to a Synthetic Case Study

We illustrate the application of this simultaneous geostatistical inversion on part of the Stanford VI-E synthetic dataset (Lee and Mukerji 2013). Only layers 1 and 2 of the original dataset were used in this example. These layers correspond to non-stationary sinuous and meandering channels.

The available synthetic dataset was comprised of 3D volumes for the measured specific resistance and seismic reflection data as well as the original petro-elastic models from where the geophysical response was modelled (Fig. 6.2). A set of 32 wells with saturation and porosity logs (Fig. 6.2) and a pre-calibrated rock physics model (see Lee and Mukerji 2013) linking the petrophysical with the elastic properties were also considered as input data for the simultaneous inversion. The reservoir grid has $150 \times 200 \times 120$ cells in the i, j and k directions, respectively.

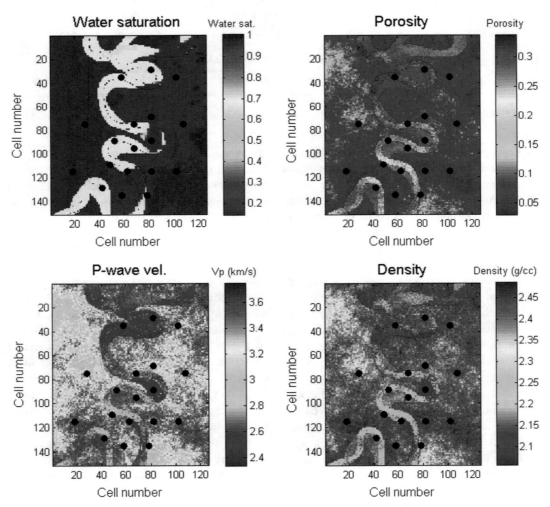

Fig. 6.2 Horizontal slices extracted at the same depth from the original 3D petro-elastic models of Sw, porosity, P-wave velocity and density and well locations for the available well-log data used to constrain the inversion

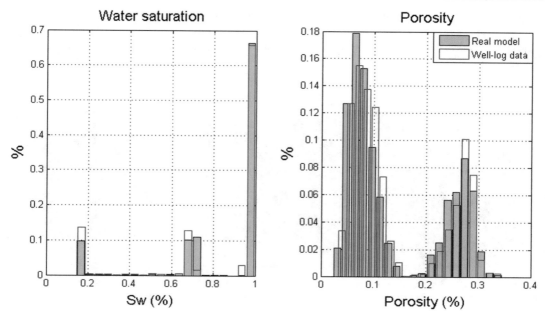

Fig. 6.3 Comparison between the original well-log data before and after the upscaling

Table 6.1 Simplified rock physics model used to link the elastic and the petrophysical domains within the iterative simultaneous geostatistical inversion (adapted from Lee and Mukerji 2013)

Mineral	Fraction (%)	Bulk modulus(GPa)	Shear modulus (GPa)	Density (g/cc)
Quartz	0.65	36.6	44	2.65
Clay	0	21	21	2.5
Feldspar	0.2	75.6	75.6	2.63
Rock fragments	0.15	80	80	2.7

Each cell is 25m by 25 m by 1 m, resulting in a reservoir size of 3750 m by 5000 m.

Since the original resistivity and seismic reflection data are noise free, some noise was introduced by upscaling the original well-log data, with very high vertical resolution, into the reservoir grid. The upscaling technique ensures the mean and variance, as estimated from the original data, are reproduced in the upscaled cells (Fig. 6.3). In addition, and in order to increase the complexity of this synthetic example, the rock physics model used as part of the geostatistical seismic inversion was also simplified in comparison to the original rock physics model as described in Lee and Mukerji (2013) (Table 6.1). The original rock physics model was built applying a constant cement model and Gardener's law (Mavko et al. 2003).

Due to the synthetic nature of the example shown here, the weights defined by Eq. 6.3 were both set to 0.5, ensuring both geophysical data have the same importance during the geostatistical inversion procedure.

Another considerable difference with respect to the original synthetic dataset is the way the synthetic specific resistance is calculated within the joint inversion workflow. While in the original synthetic dataset real specific resistance was calculated using Archie's Law (Archie 1942) for sand facies and Waxman and Smits (1968) for shales, we oversimplified this calculation by using Archie's Law for the entire model without including any facies model within the joint inversion methodology. The Archie's parameters used to calculate the synthetic specific resistance are synthetized in Table 6.2.

Table 6.2 Archie's parameters used as part of the geostatistical joint inversion workflow (adapted from Lee and Mukerji 2013)

Parameter	Tortuosity constant (a)	Cementation exponent (m)	Saturation exponent (n)	Brine resistivity (Ω m)
Value	1	2	1.8	0.25

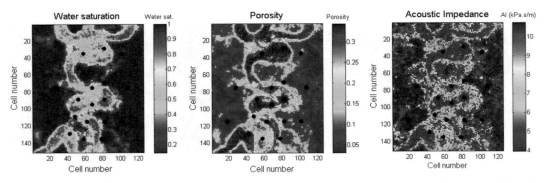

Fig. 6.4 Horizontal slices extracted at the same depth as shown in Fig. 6.2 from the mean model calculated from the petro-elastic models generated during the last iteration of: (from *left*) water saturation, porosity and acoustic impedance

The results shown here were obtained after six iterations, in which at each iteration a set of 32 pairs of Sw and porosity models were simulated by DSS and co-DSS with joint probability distributions. At the end of the iteration the average global correlation coefficient between the entire volume of real and synthetic resistivity and seismic reflection data is about 0.75.

The mean model calculated from the set of realizations generated during the last iteration produces highly-correlated synthetic seismic and EM data for all realizations. For this case study, during the last iteration the 32 realizations of Sw and porosity produced synthetic seismic reflection and resistivity data with a good match when compared to the real responses: i.e. global correlation coefficients above 0.7. Therefore, the petro-elastic generated at this iteration only show small-scale variability and its mean model is a good approximation of the convergence level for the geostatistical procedure.

The mean model of the petro-elastic models generated during the last iteration of Sw and porosity have a good match for the real ones (Fig. 6.4). The inverted models show more discontinuities within the meandering channels compared with real petro-elastic models;

nevertheless, the shape and spatial location of the main sedimentary structures, along with their spatial distribution, is well reproduced. The small-scale variability within the channel areas is a direct result of the stochastic approach used to solve this inversion problem. The proposed joint inverse methodology can retrieve high resolution petro-elastic models. Another important aspect worth mentioning is the underestimation of the values of the inverted properties of interest outside the channel areas.

Finally, the synthetic resistivity and seismic responses from the mean models presented in Fig. 1.5 agree with the real recorded data (Fig. 6.5). It is clear, however, that the match is considerably better for the seismic reflection data when compared to the data for specific resistance. The specific resistance values outside the channels are overestimated in comparison to the real data.

It should be noted that this synthetic dataset is highly non-stationary and suitable for geostatistical methodologies approaches based on multi-point statistics (e.g. Strebelle 2002). However, while here we use stochastic sequential simulation algorithms based on bi-point statistics, the proposed geostatistical joint inverse methodology succeeds

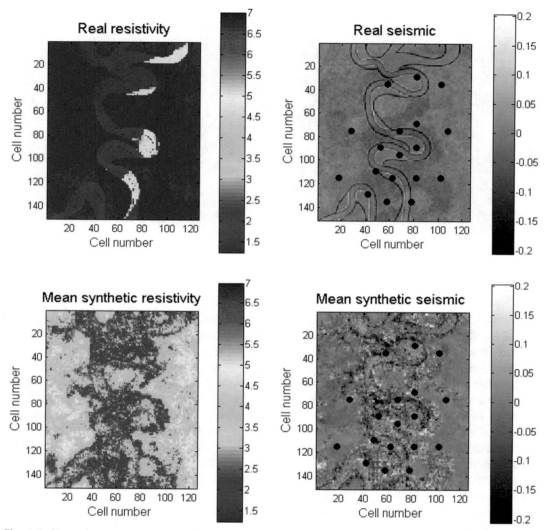

Fig. 6.5 Comparison between horizontal slices extracted at the same depth as shown in Fig. 6.8 from (*top*) real specific resistance and seismic reflection data and (*bottom*) synthetic specific resistance and seismic response calculated from the mean model resulting from the petro-elastic models generated during the last iteration (Fig. 6.2)

in reproducing the main spatial features of the complex geological environment.

6.1.1.1   Final Remarks About Joint Inversion of Seismic and CSEM Data

The results obtained seem robust, even in challenging non-stationary environments such as the one represented by this synthetic dataset. The retrieved inverse water saturation, porosity and

acoustic impedance (or density and P-wave velocity) models reproduce both small-scale and large-scale details as interpreted from the real models.

The petro-elastic properties best inferred from the geostatistical joint inverse workflow are porosity and acoustic impedance (or P-wave velocity and density). On the other hand, Sw is the property in which the algorithm shows lower convergence towards the solution. The results do

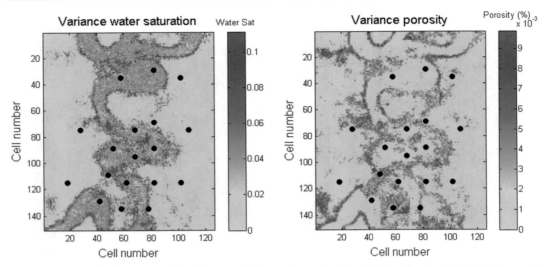

Fig. 6.6 Horizontal slices extracted at the same depth as shown in Fig. 6.2 from the standard deviation model calculated from the ensemble of petro-elastic models generated during the last iteration

agree with what it is expected, since the resolution of the EM data is smaller in comparison to the seismic reflection data.

The inverted petro-elastic models have a lower speed of convergence outside the channel areas in comparison to the real ones. We put this mismatch of property values down to two reasons: the difference between the distributions of the real three-dimensional models in comparison to those retrieved from the upscaling data; and the simplifications in the calculation of the specific resistance response. While the high specific resistance areas within the channel are fairly well reproduced, there is a clear mismatch (overestimation) of resistivity values. It is clear to us that the proposed methodology should include a facies model when computing the synthetic resistivity response from the Sw and porosity models. The over-simplification of modeling the synthetic resistivity response allowed us to test and assess the concept of the use of a geostatistical framework for jointly inverting geophysical data of very different nature.

As an iterative geostatistical joint inversion methodology, the spatial uncertainty of each property may be assessed individually by calculating the variance between the set of petro-elastic models simulated and co-simulated during the last

iteration of the inversion procedure (Fig. 6.6). The interpretation of the variance models shows the porosity models are related with the location and spatial extent of the meandering channels, while the Sw models are related with the fluid fill within the channel itself. This is directly related to the nature of the geophysical data integrated within this inversion methodology. The seismic reflection data allowing the inversion directly for porosity has a stronger relationship with the location and structure of the study area, while the EM data is highly responsive to the pore fluids present in the subsurface rocks.

6.2 Integration of Dynamic Production Data: Global Inversion

Here we present a contribution to one of the most important and difficult challenges of the hydrocarbon reservoir characterization: the integration of all the available geological, seismic reflection, well-log and production data into a coherent numerical reservoir model to reflect the Earth's subsurface complexity and heterogeneities.

Traditional approaches within the oil and gas industry for reservoir characterization is still heavily based on matching each type of data sequentially in separate workflows. Subsurface elastic, or petro-elastic, models derived from any inversion procedure, as, for example, those described above, are able to match the observed seismic data. However, these models are frequently unsuitable for matching and predicting production data in mature fields. In such cases, the inverted petro-elastic models are perturbed in order to match the historical production data in a process commonly designated as history matching (Hu 2000; Hu et al. 2001; Hoffman and Caers 2005; Caers and Hoffman 2006; Kashib and Srinivasan 2006; Mata-Lima 2008; Oliver et al. 2008; Demyanov et al. 2011; Oliver and Chen 2001).

History matching is another inverse problem in which the known solution, the measured observations of production ($\mathbf{d}$), is related with the unknown parameters ($\mathbf{m}$) through a nonlinear function ($\mathbf{g}$), e.g. a fluid flow simulator:

$$\mathbf{d} = \mathbf{g}(\mathbf{m}) + \mathbf{e}, \qquad (6.4)$$

In history matching problems we try to match the reservoir's response at sparse well locations, but allowing modifications for the entire reservoir grid without spatial constraint from other data (for example, seismic data). There is a highly nonlinear relationship to optimize between the local petrophysical properties at the well locations and the model parameters. In addition, there is also a spatial scale gap between the petrophysical properties one tries to infer spatially-distributed along the entire reservoir field and the dynamic responses obtained locally at the well locations. Different history matching procedures are proposed to solve this problem by using geostatistics as the driving process of parameter characterization (Hu 2000; Hu et al. 2001; Hoffman and Caers 2005; Caers and Hoffman 2006; Kashib and Srinivasan 2006; Mata-Lima 2008). Essentially, these methods consist of an iterative procedure with the perturbation of the model parameter space by stochastic sequential simulation and an optimization process to guarantee the convergence

through the desired solution: i.e. the known production data. By tuning the inferred reservoir models to match historical production data, the resulting petro-elastic models generally begin diverging from the observed seismic reflection data, particularly at locations far from the wells, where there is no constraining 'hard-data'.

The next section introduces traditional geostatistical history matching, as proposed by Mata-Lima (2008). It is then followed by an approach that simultaneously integrates seismic reflection and production data within the same history matching iterative procedure. Note that despite the distinct physical nature of these inverse problems (seismic inversion and history matching) both have the same parameter solution space: the reservoir's model parameter space.

6.2.1 Geostatistical History Matching

The traditional geostatistical history matching (Mata-Lima 2008; Caeiro et al. 2015) is an iterative geostatistical procedure that uses stochastic sequential simulation as the model perturbation technique and a genetic algorithm as a global optimizer for the iterative procedure. It uses available well-log data as experimental data for the geostatistical simulation and a variogram model to describe the spatial continuity pattern of the property being modelled. Within this framework, there is the assumption of stationarity for the first and second statistical moments as estimated from the available well-log data. We assume the spatial dispersion behavior of the natural property we are seeking to model can be fully described by a variogram model for the entire reservoir grid. It is true that for highly non-stationary geology settings, such as those associated with turbidite channels, these assumptions generally do not meet realistic geological models; however, given the simplicity of the perturbation technique of the model parameter space they allow a faster convergence in the iterative processes. For more on non-stationary inverse problems, please see Sect. 4.3.8 in Chap. 4.

The traditional geostatistical history matching may be summarized through the following sequence of steps (Mata-Lima 2008):

(1) Simulate an ensemble of petrophysical models (e.g. porosity and/or permeability) from the available well-log data for the entire reservoir grid using a secondary model derived from a seismic inversion procedure;

(2) Complete a fluid flow simulation for each of the models generated in (1);

(3) Compare the production data obtained in (2) and the real historical production data;

(4) Divide the reservoir into areas of influence for each well being considered;

(5) For each area considered in the previous step, select the petrophysical model that ensures the maximum match with the production data for that particular area;

(6) Use the patchwork model composed in (5) as secondary variable for the generation of a new ensemble of models using stochastic sequential co-simulation;

(7) Iterate until all wells match the observed production data.

The main bottlenecks of this family of history matching algorithms are the considerable amount of computation time spent on the fluid flow simulations (one for each model generated during the entire iterative procedure) and the lack of a spatial conditioning data for cells located far from the wells. The only spatial constraint at these locations is the global cumulative distribution function of the simulated property and the variogram model imposed during the geostatistical simulations.

6.2.2 Iterative Global Seismic Inversion in History Matching

The integration of seismic reflection and production data into the traditional geostatistical history matching allows for more detailed and constrained subsurface Earth models. While the detail increases, their intrinsic uncertainty is reduced. Hence, geostatistical seismic inversion (Sect. 4.3 in Chap. 4) and geostatistical history matching are two different inversion methods with the same aim—a numerical model of parameters—and identical perturbation in an iterative procedure—DSS and co-simulation with joint probability distributions—the solution of both inverse problems cannot be the same. The idea of the proposed approach is to obtain a solution in an iterative process that jointly matches both objective functions: the match between synthetic and real seismic data and between historical and simulated production data. The integration of seismic reflection data into the geostatistical history matching workflow may be split into two stages: the stochastic simulation of petro-elastic models—forward modeling and the comparison against the observed seismic and production data (Fig. 6.7)—and the selection of the conditioning data for the next iterations based on the petro-elastic ensemble simulated at the current iteration (Fig. 6.8).

The first stage of this simultaneous inversion procedure may be further divided into two processes that run in parallel: the geostatistical seismic inversion of post-stack seismic data (as described in Sect. 4.4.3 in Chap. 4) and the fluid flow simulation and comparison against the observed data (Fig. 6.7).

First, a global geostatistical acoustic inversion, as described in Sect. 4.3.4 in Chap. 4, is performed. For each simulated Ip model, Ns porosity is generated by stochastic sequential simulation with joint probability distributions from where Ns permeability distributions are derived following the same methodology. In this way, the relationship between the petro-elastic properties is ensured between the simulated and co-simulated models as retrieved from the available well-log data.

From the set of Ns acoustic impedance models, Ns post-stack synthetic seismic volumes are calculated. Each synthetic volume is then compared, in terms of correlation coefficient, with the real seismic data on a trace-by-trace basis. These correlation coefficients can be thought as seismic

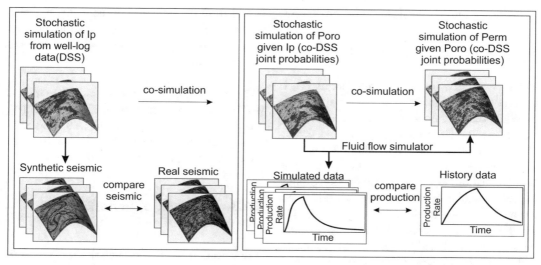

Fig. 6.7 Schematic representation of the first stage of the geostatistical history matching conditioned to seismic inversion workflow

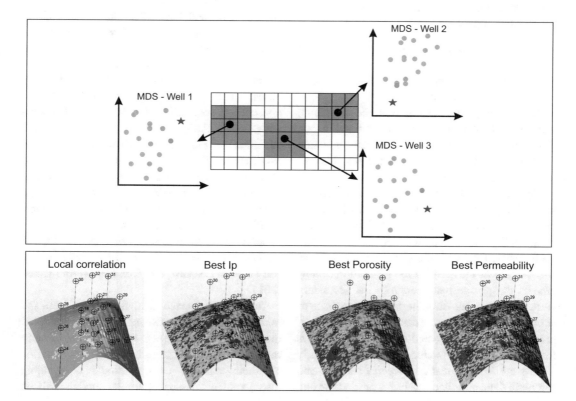

Fig. 6.8 *top* MDS plot for three wells with 16 models each (*blue circles*). The area of influence of each well is represented in *orange* over a 2D grid. The model plotted with a *green circle* corresponds to the model with the lowest mismatch in terms of simulated and observed production data and will be used to fill the respective area of influence in the best local correlation and best cubes. *bottom* Best local correlation cube and best volumes for acoustic impedance, porosity and permeability. The grid cells outside the areas of influence are filled with the best correlated models in terms of seismic data

deviations from the real or recorded seismic data:

$$e_s^l = \left(A^l - A_r\right), \quad l = 1, \ldots, N_s, \qquad (6.5)$$

where A^l is the synthetic seismic data and A_r is the real seismic data.

While the seismic forward modeling is being performed, the Ns pairs of porosity and permeability are used as input for a fluid flow simulator (e.g. Eclipse®, T-Navigator®). From the fluid flow simulation, we obtain Ns production profiles for the variables of interest (e.g. oil and water production). Depending on the target of the history matching (e.g. total field or well production rates), the simulated responses are compared with the corresponding historical production data: the observed data. For each simulated production profile, Ns deviations from the observed historic production data can be calculated following:

$$e_d^l = \left(d^l - d_r\right), \quad l = 1, \ldots, N_s, \qquad (6.6)$$

where d^l is the synthetic production profile and d_r is the historic production data.

At this stage we can assess the mismatch between synthetic and real seismic reflection data (Eq. 6.5) and the deviations between simulated fluid flow production data and the real historic production data (Eq. 6.6).

The second step of the proposed iterative methodology comprises the selection of the best models of acoustic impedance, porosity and permeability, and the best local correlation coefficients at the end of a given iteration. These will be used as the seed for the generation of new models during the next iteration, i.e. secondary variables in the co-simulation of a new set of petro-elastic properties. These are composite, or patchwork, models created by selecting patches from the set of simulated models that locally ensure the lowest misfit between observed and synthetic data, simultaneously for production and seismic reflection data. The proposed methodology is based on the following rational:

- for the cells close to the well locations, the misfit of production data will prevail in choosing the best models of the next iteration;

- far from the influence of the wells, the misfit between synthetic and real seismic reflection data will condition the choice of the best models for the next iteration.

The selection of the areas of influence may be performed strictly by statistical algorithms (e.g. Voronoi polygons), based on well tests or a geological facies model that is inferred, for example, from previous inverted models.

6.2.3 Selection of Petro-Elastic Models

The selection of the petro-elastic models to fill the areas of influence for each well is carried out by plotting the simulated productions along with the real ones in a multidimensional scaling referential (Cox and Cox 1994; Scheidt and Caers 2008; Suzuki and Caers 2008; Caers 2011). Multidimensional scaling (MDS) is a multivariate statistical technique that can reveal, in few dimensions, the patterns between a set of multidimensional models based on the concept of distances (Caers 2011). Briefly, the MDS converts a dissimilarity matrix (D) into points, which can then be plotted in a Cartesian space, the MDS space (Borg and Groenen 1997; Cox and Cox 1994; Caers 2011). The matrix D is first converted into a matrix A by a scalar product. Then, A is decomposed by eigenvector decomposition, where only the first d principal components, or eigenvectors, are retained. A detailed mathematical description of this methodology can be found in Cox and Cox (1994) and Caers (2011).

In the metric space, the MDS space itself, the relative position between several simulated models or their dynamic responses, is directly related to how similar these models are in terms of their internal configuration or dynamic responses. In this space, similar models will be plotted in a cluster, while distinct models will be plotted with greater distances between them.

An essential step in the MDS procedure is the selection of the distance to construct the dissimilarity matrix (D). In order to ensure a good separation of the ensemble of models in the MDS space, this distance should reflect the type of data one is dealing with. If the goal is to distinguish between models that have bodies with different shapes (e.g. different channelized systems), then the Hausdorff distance may be suitable for that purpose (Scheidt and Caers 2008; Suzuki and Caers 2008). Correlation-based distances are preferred for distinguishing between different temporal signals such as seismic reflection data. For the geostatistical seismic inversion with history matching we calculate the Euclidean distance per well between each of the simulated production curves against its corresponding real curve.

After calculating the distance matrix (D) for each individual well, all the simulated responses and historic data are plotted in the MDS space (Fig. 6.8). Then, for each well in the corresponding metric space, the petro-elastic models with the production profile that is closest to the observed production data will be chosen to build the best acoustic impedance, porosity and permeability models within the area of influence for that particular well. The associated local correlation coefficients calculated from the synthetic seismic traces resulting from the selected elastic models and the real seismic data are also stored in a local correlation cube.

All the other zones outside the well' area of influence are selected according the match against the recorded seismic data: i.e. following the traditional approach of the geostatistical seismic inversion methods introduced in Chap. 3.

The geostatistical iterative procedure continues with the generation of a new set of petro-elastic models recurring to co-simulation and using the best acoustic impedance, porosity and permeability models as secondary variables, along with the best local correlation volume. The inversion is considered complete when the global correlation coefficient between a synthetic seismic volume and the real seismic data is above a certain threshold. This simultaneous inversion methodology can be summarized through the following sequence of steps (Fig. 6.9):

(1) Stochastic sequential simulation of Ns acoustic impedance models using DSS;

(2) Stochastic sequential co-simulation of Ns porosity models with co-DSS with joint probability distributions using the models simulated in (1) as secondary variables;

3) Stochastic sequential co-simulation of Ns permeability models with co-DSS with joint probability distributions using the models simulated in (2) as secondary variables;

(4) Calculate Ns synthetic seismic volumes and compare each with the real seismic data on a trace-by-trace basis for each model simulated in (1);

(5) Calculate the respective flow simulations and compare the obtained dynamic responses with the observed ones from the historic production data for each pair of porosity and permeability in (2) and (3);

(6) Plot the simulated and real production profiles in the MDS space for each well individually;

(7) Compose new best acoustic impedance, porosity and permeability models from the set of simulated models in (1), (2) and (3). Around the well location select the models from (6) with production data closest to the real one. Select the models that produce the synthetic seismic data that best correlates with the real one in areas far from the influence of the wells;

(8) Based on a global genetic algorithm, use the 'best' models created in (7) as secondary variables in the perturbation of the model parameters by using co-simulation methodology. Return to (1) and iterate until the global correlation coefficient between real and synthetic seismic data is above a certain threshold.

As a geostatistical approach, all simulated models created during each iteration can reproduce the probability distributions for the inverted properties, acoustic impedance, porosity and permeability, as estimated from the well-log data; the joint distribution between acoustic impedance versus porosity and porosity versus permeability as retrieved from the well-log data; the spatial continuity pattern imposed by the variogram; and the values of the well-logs at the well locations.

The inverted petro-elastic models are able to simultaneously match the observed seismic and

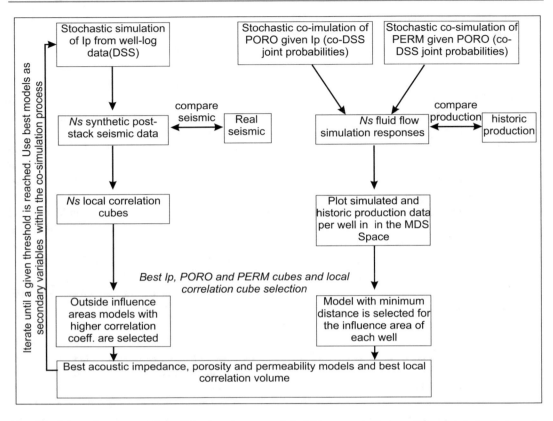

Fig. 6.9 Schematic representation of the iterative geostatistical history matching conditioned to seismic inversion workflow

historic production data. The proposed approach solves two different highly nonlinear inverse problems within the same solution space: the model parameter space. From the resulting set of inverted models, the individual spatial uncertainty of each property can also be assessed, allowing for better risk assessment.

The main bottleneck of geostatistical history matching methodologies is related to the computational burden associated with the Ns fluid flow simulations carried out at each iteration in order to obtain the simulated production data for each pair of porosity and permeability models. For this reason, in medium to large reservoir models, the proposed method can accomplish the fluid flow simulation in a coarser reservoir grid in a multi-scale approach (Marques et al. 2015). These alternative solutions can be easily coupled

with the iterative geostatistical history matching conditioned by seismic inversion.

6.2.4 Application to a Synthetic Case Study

In this section we show the application of both methodologies described above to part of the Stanford VI synthetic reservoir (SVI; Castro et al. 2005). A portion of $60 \times 75 \times 20$ cells in the i, j, k directions, respectively, was selected from layer 2 of the SVI (Fig. 6.10). The original reservoir grid was upscaled into a coarser grid comprising 50 m $\times$ 50 m $\times$ 2 m cells in order to speed up the fluid flow simulator. The upscaling procedure ensured the reproduction of the original spatial distribution for the petro-elastic models as well as

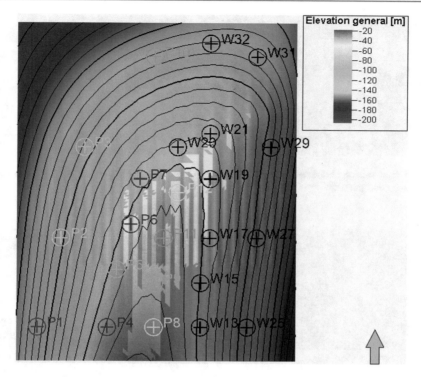

Fig. 6.10 Available set of wells and their location within the study area. Coloured wells were used to constrain the geostatistical inversion while black wells were used exclusively as blind tests. Top reservoir surface is also shown as example

its distribution functions. Only 12 of the 23 wells were used as a constraint for the geostatistical history matching and seismic inversion (Fig. 6.10). The remaining wells were not used in any part of the inversion procedure and were kept exclusively for blind tests at the well locations. The rest of the available dataset comprises real three-dimensional models of acoustic impedance, porosity and permeability (Fig. 6.11) as well as a noise-free full-stack seismic volume (Fig. 6.12).

Structurally, the SVI reservoir is a simple and gentle asymmetrical anticline with axis N15E°. Layer 2 of the SVI dataset comprises meandering channels of variable sizes with four facies types: floodplain, point bar, channel and boundaries (Castro et al. 2005). The reproduction of such a non-stationary sedimentary environment in reservoir modeling represents a challenge for any geostatistical inversion methodology based on two-point geostatistics. The shape and thickness of the many meandering channels vary

considerably across the real petro-elastic models (Figs. 6.11 and 6.12) and a successful inversion needs to ensure the reproduction of these channels in terms of size and spatial distribution.

Finally, a fluid flow simulator, such as *Eclipse®* (Schlumberger) or T-Navigator (RFD), ran over the original petrophysical models to produce the historic data used for the history matching problem. The reservoir was in production for approximately three years. The simulation target was bottom-hole pressure (BHP) at each individual well. The wells were shut down after reaching a minimum BHP value. An oil production rate (OPR) constraint was also applied to each individual well. The initial reservoir conditions were set such as not all 12 wells were producing simultaneously during the three years. The intrinsic parameterization of the initial reservoir conditions is not an important aspect in testing the performance and efficiency of the integration of seismic data into history matching.

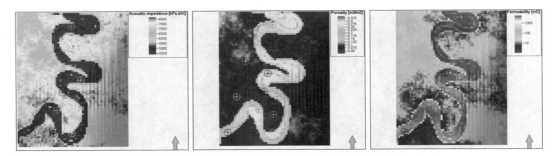

Fig. 6.11 Horizontal sections extracted from the real three-dimensional petro-elastic models. From *left*: acoustic impedance, porosity and permeability

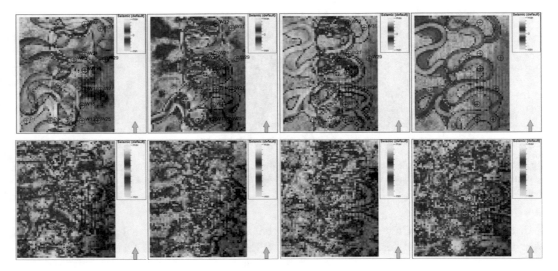

Fig. 6.12 Horizontal sections extracted from (*top*) the real seismic data and (*bottom*) the best-fit synthetic seismic data at the end of the geostatistical inversion at different depths. The inverted seismic data provides a good reproduction of the sedimentary bodies present within the original data

In this particular example, the geostatistical history matching conditioned to seismic inversion converged to a global correlation coefficient between real and synthetic seismic data of 0.81. This convergence was obtained after six iterations with 16 ensembles of petro-elastic models (acoustic impedance and porosity) simulated and co-simulated per iteration. The resulting synthetic seismic data with the highest correlation coefficient matches the main seismic reflections as observed in the real seismic data (Fig. 6.12). The integration of both data allows the reproduction of both the non-stationary patterns related with the meandering channels and the high

variability that they present in terms of shape and thickness.

The resulting inverted petro-elastic models for acoustic impedance, porosity and permeability (Figs. 6.13, 6.14 and 6.15) reproduce the real ones very well. The inverted models can reproduce the small-scale and large-scale non-stationary patterns. It is also interesting to note the evolution of the inverted models from iteration to iteration (Fig. 6.13, 6.14 and 6.15). In the first iteration, the simulated models are conditioned only to the available well-log data and, consequently, the meandering structures are not reproduced. After the first iteration, the inverted

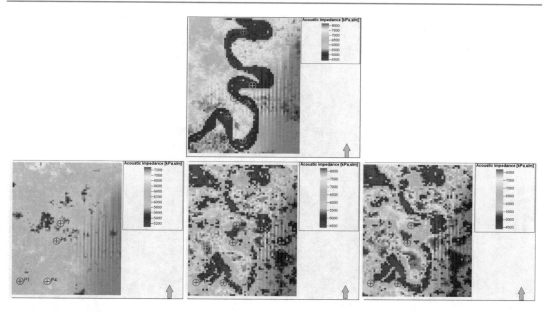

Fig. 6.13 Horizontal sections extracted from (*top*) real Ip model and (*bottom*) the mean of the acoustic impedance models simulated at: (from *left* to *right*) iteration 1, iteration 3 and iteration 6

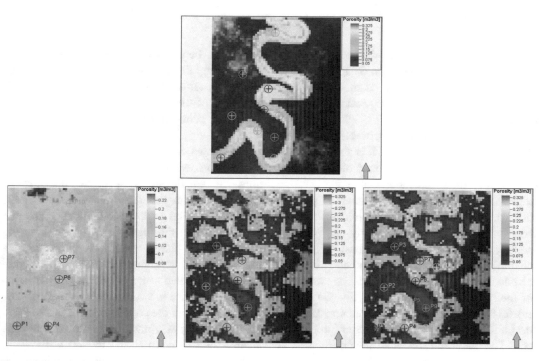

Fig. 6.14 Horizontal sections extracted from (*top*) real porosity model and (*bottom*) the mean of the porosity models simulated at: (from *left* to *right*) iteration 1, iteration 3 and iteration 6

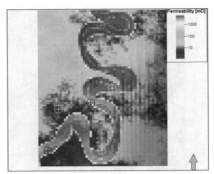

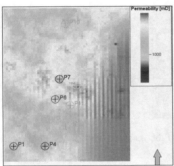
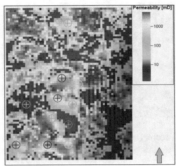
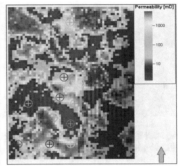

Fig. 6.15 Horizontal sections extracted from (*top*) the real permeability model and (*bottom*) the mean of the permeability models simulated at: (from *left* to *right*) iteration 1, iteration 3 and iteration 6

petro-elastic models begin conditioned to the real seismic and historic production data, in this way beginning to reproduce the main sedimentary features as interpreted from the real petro-elastic models.

Along with the inverted petro-elastic models and the synthetic seismic data, it is also necessary to assess the production responses obtained from each of the converged models after the iterative geostatistical procedure. The production profiles for each well, obtained by the fluid flow simulation of the inverted permeability and porosity models, and related with the highest global correlation coefficient between synthetic and real seismic reflection data, are impressive matches for the historic production data (Fig. 6.16, 6.17 and 6.18). Generally, history matching is better for OPR and BHP than for water production ratio. The well with the worst match in terms of water production rate is P3. It is worth noting the petro-elastic models designated by best-fit models (plotted in red in Fig. 6.13, 6.14 and 6.15) are those producing the greatest global correlation

coefficient between real and synthetic seismic data during the iterative inversion process. In fact, for most of the wells, these best-fit models also present the best match between the set of simulated petro-elastic models and the historic data.

As shown in Fig. 6.13, 6.14 and 6.15, the best-fit inverted petro-elastic models can reproduce the small- and large-scale structures as interpreted from the real petro-elastic models. The reproduction of the non-stationary patterns associated with the meandering channels in the inverted petro-elastic models is surprising. In addition to the reproduction of the spatial patterns, the inverted petro-elastic models are also able to reproduce the values of the real property. In order to assess the convergence of the inverted models locally, we compared the inverted models against the real petro-elastic models at the locations of wells not used to constrain the inversion (Fig. 6.19). The reproduction of the original well-logs for acoustic impedance, porosity and permeability is extremely good, even for

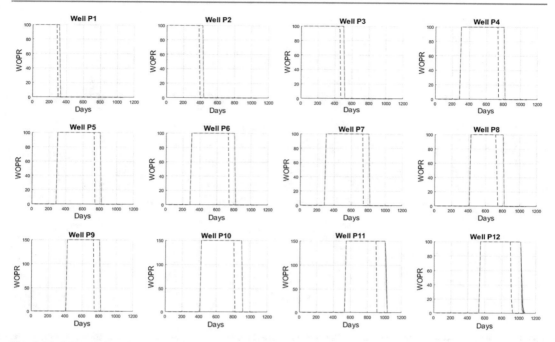

Fig. 6.16 OPR profiles for the best-fit inverted model for each well individually during the three-year production period. The *black dashed curve* is the historic production data. The *red dashed line* corresponds to the response of the best-fit inverted models. The fluid flow responses of all models simulated during the last iteration are plotted behind the *thick red line*

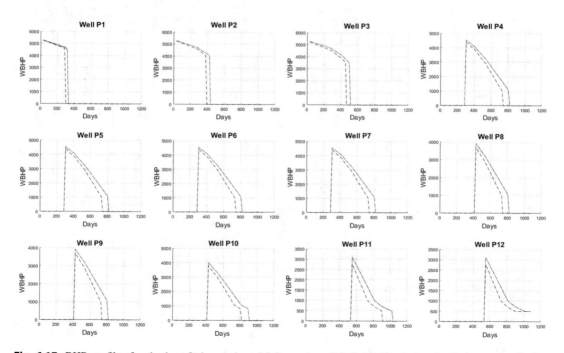

Fig. 6.17 BHP profiles for the best-fit inverted model for each well individually during the three-year production period. The *black dashed curve* is the historic production data. The *red dashed line* corresponds to the response of the best-fit inverted models. The fluid flow responses of all models simulated during the last iteration are plotted behind the *thick red line*

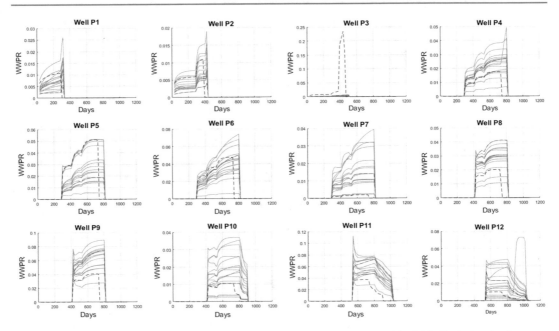

Fig. 6.18 Water production rate profiles for the best-fit inverted model for each well individually during the three-year production period. The *black dashed curve* is the historic production data. The *red dashed line* corresponds to the response of the best-fit inverted models and thin *grey lines* to the fluid flow responses of all models simulated during the last iteration

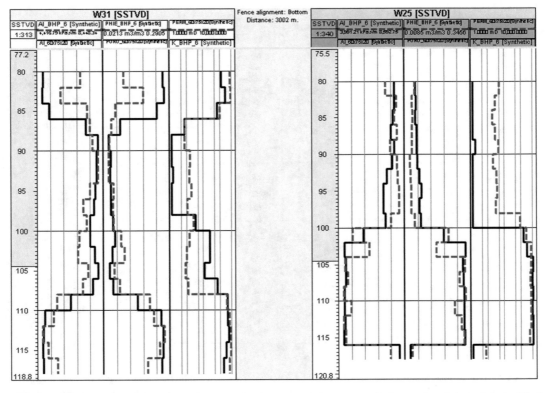

Fig. 6.19 Blind well tests for W31 and W25 well (for location see Fig. 6.14). The inverted models (*red dashed line*) match the real ones (*black solid line*) at these locations. From *left*: acoustic impedance, porosity and permeability. The fit is particularly good for acoustic impedance and porosity

(a) **(b)**

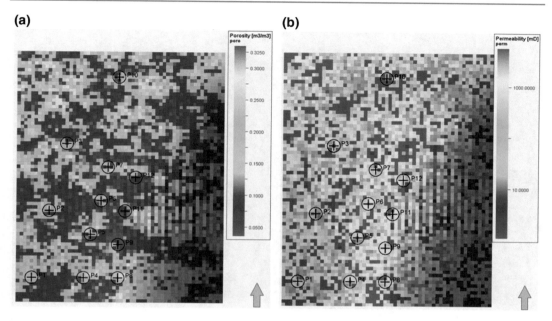

Fig. 6.20 Horizontal sections extracted from the best-fit inverse models of **a** porosity and **b** permeability retrieved from the traditional geostatistical history matching

positions far from the main cluster of wells, and consequently less constrained by the experimental data, such as the case of wells W31 and W25.

To reveal the importance of integrating dynamic production data within the same inversion procedure, we show the results of applying the traditional geostatistical history matching (as described in Mata-Lima 2008, Sect. 6.2.1) to the non-stationary dataset used in the previous example.

Horizontal sections extracted from the best-fit pair of porosity and permeability models after six iterations of the iterative geostatistical history matching procedure are shown in Fig. 6.20. The a priori configuration for this example is the same as that shown from the integration of seismic reflection data into history matching. It is clear the traditional geostatistical history matching fails to retrieve a channelized configuration exclusively from the well-log and production data. This is mainly due to a lack of spatial constraints in the productive areas. The incorporation of seismic reflection data within the

history matching procedure can fill this gap, allowing for more reliable petrophysical models.

6.2.5 Final Remarks

The use of bi-point statistics, such as the variogram, to reproduce production data in traditional geostatistical history matching usually fails to reproduce non-stationary channelized structures. Multi-point statistics, with appropriated training images, succeed in reproducing realistic geological models of such structures, but usually fail to integrate the production data once there is a lack of malleability to perturb models in iterative steps while preserving the channelized features of interest.

This proposed approach of integrating seismic reflection data in history matching seems to be promising and an efficient way of coupling both objectives: reproducing the production data and the main geological features such as channelized structures as interpreted from the available seismic reflection data.

This book seeks to fill the gap between traditional geostatistical methodology tools for reservoir modeling and characterization, and inverse procedures for integrating different data within the geo-modeling workflow. It begins with a review of stochastic sequential simulation. The geostatistical inversion algorithms introduced here were presented to be used as a first approach by anyone, from student to geoscientists, generating reservoir models on a daily basis.

We hope to have aroused curiosity to engage in further research and embark in new developments in important fields, ranging from petrophysics to reservoir engineering. We believe the new and challenging environments, such as deep offshore, require integrative and multidisciplinary approaches that incorporate knowledge from different scientific areas. Moreover, due to their complexity we should never forget uncertainty that is always present at every stage—from data processing to modeling and characterization. This uncertainty should be assessed at all stages and integrated within entire production chain. Improved uncertainty assessment allows for better decision making with fewer risks and greater success. We hope that we have been able to show that the methodologies presented here can help achieve this goal.

© Springer International Publishing AG 2017
L. Azevedo and A. Soares, *Geostatistical Methods for Reservoir Geophysics*,
Advances in Oil and Gas Exploration & Production, DOI 10.1007/978-3-319-53201-1_7

References

Abrahamsen, P., Hauge, R., & Kolbjornsen, O. (Eds.). (2012). *Geostatistics Oslo 2012*. Netherlands: Springer.

Alabert, F. G. (1987). *Stochastic imaging of spatial distributions using hard and soft information* (MSc Thesis). Stanford University, CA.

Almeida, A. S., & Journel, A. G. (1994). Joint simulation of multiple variables with a Markov-type coregionalization model. *Math Geology, 26,* 565–588.

Anderson, T. W. (1984). *An introduction to multivariate statistical analysis*. London: Wiley Inc.

Armstrong M. (Ed.). (1989). *Geostatistics* (Vol. 1 and 2). Dordrecht: Kluwer Academic Publishers.

Armstrong, M., & Dowd, P. (Eds.). (1994). Geostatistical simulations. In *Proceedings of a workshop, Fontainebleau, France*, May 27–28, 1993. Kluwer Academic Press, Dordrecht, Holland.

Archie, G. E. (1942). The electrical resistivity log as an aid in determining some reservoir characteristics. *Petroleum Transactions of AIME, 146,* 54–62.

Arpat, G., & Caers, J. (2007). Conditional simulation with patterns. *Mathematical Geology, 39*(2), 177–203.

Avseth, P., Mukerji, T., & Mavko, G. (2005). *Quantitative seismic interpretation*. Cambridge: Cambridge University Press.

Azevedo, L., & Soares, A. (2014). Geostatistical joint inversion of seismic and electromagnetic data. *Geosciencias Aplicadas LatinoAmerica, 1,* 45–52.

Azevedo, L., Nunes, R., Soares, A., & Neto, G. S. (2013). Stochastic seismic AVO inversion. In *75th EAGE Conference & Exhibition*, June 2013: 10–13.

Azevedo, L., Nunes, R., Soares, A., Mundin, E. C., & Neto, G. S. (2015). Integration of well data into geostatistical seismic amplitude variation with angle inversion for facies estimation. *Geophysics, 80*(6), M113–M128.

Baafi, E.Y., & Schofield, N. A. (Eds.). (1997). *Geostatistics Wollongong '96* (Vol. 1 and 2). Dordrecht: Kluwer Academic Publishers.

Borg, I., & Groenen, G. (1997). *Modern multidimensional scaling: Theory and applications*. Netherlands, New York: Springer.

Borgman, L., Taheri, M., & Hagan, R. (1984). Three-dimensional, frequency-domain simulations of geological variables. In G. Verly, M. David, A. G. Journel, A. Marechal (Eds.), *Geostatistics for natural resources characterization*, Dordrecht, Holland: D. Reidel Publishing.

Bortoli, L. J., Alabert, F., Haas, A., & Journel, A. G. (1993). Constraining stochastic images to seismic data. In A. Soares (Ed.), *Geostatistics Troia'92* (pp. 325–337). Dordrecht: Kluwer.

Bosch, M., Mukerji, T., & González, E. F. (2010). Seismic inversion for reservoir properties combining statistical rock physics and geostatistics: A review. *Geophysics 75*(5): 75A165. doi:10.1190/1.3478209

Boschetti, F., Dentith, M.C., & List, R. D. (1996). Inversion of seismic refraction data using genetic algorithms. *Geophysics,61*(6) (November), 1715–1727. doi:10.1190/1.1444089.

Boucher, A., & Dimitrakopoulos, R. (2007). Block simulation of multiple correlated variables. *Mathematical Geosciences,41*(2), 215–237.

Bourgault, G., & Marcotte, D. (1991). Multivarite variogram and its application to the linear model of corregionalization. *Mathematical Geology, 23*(7), 899–928.

Buland, A., & Omre, H. (2003). Bayesian linearized AVO inversion. *Geophysics, 68*(1), 185–198.

Buland, A., & El Ouair, Y. (2006). Bayesian time-lapse inversion. *Geophysics 71*(3): R43–R48. doi:10.1190/1.2196874

Caeiro, M. H., Demyanov, V., and Soares, A. 2015. Optimized history matching with direct sequential image transforming for non-stationary reservoirs. *Mathematical Geosciences*. doi:10.1007/s11004-015-9591-0

Caers, J. (2011). *Modeling uncertainty in earth sciences*. UK: Wiley-Blackwell.

Caers, J., & Hoffman, T. (2006). The probability perturbation method: A new look at Bayesian inverse modelling. *Mathematical Geology, 38*(1), 81–100.

Caetano, H. (2009). *Integration of seismic information in reservoir models: Global stochastic inversion* (Ph.D. Thesis). Instituto Superior Técnico, Universidade Técnica de Lisboa, Portugal

Castagna, J. P., & Backus, M. (Eds.). (1993). Offset-dependent reflectivity—theory and practice of avo analysis. *Investigations in geophysics, no. 8*. Tulsa: Society of Exploration Geophysicists

© Springer International Publishing AG 2017
L. Azevedo and A. Soares, *Geostatistical Methods for Reservoir Geophysics*,
Advances in Oil and Gas Exploration & Production, DOI 10.1007/978-3-319-53201-1

Castagna, J. P., & Swan, H. W. (1997). Principles of AVO crossplotting. *The Leading Edge, 17,* 337–342.

Castro, S., Caers, J., & Mukerji, T. (2005). *The Stanford VI reservoir. 18th Annual Report.* Stanford Center for Reservoir Forecasting (SCRF), pp. 1–73.

Chilès, J.-P., & Delfiner, P. (1999). *Geostatistics: Modeling spatial uncertainty.* New York, US: Wiley.

Coléou, T., Allo, F., Bornard, R., Hamman, J., & Caldwell, D. (2005). *Petrophysical seismic inversion.* Houston, US: SEG Annual Meeting 2005.

Contreras, A., Torres-Verdin, C., Kvien, K., Fasnacht, T., & Chesters, W. (2005). AVA stochastic inversion of pre-stack seismic data and well logs for 3D reservoir modelling. In *67th EAGE Conference & Exhibition,* June 2005: 13–16.

Cox, T. F., & Cox, M. A. A. (1994). *Multidimensional scaling.* London: Chapman and Hall.

Daly, C., & Caers, J. (2010). Multi-point geostatistics—and introductory overview. *First Break, 28*(9), 39–47.

David, M. (1977). *Geostatistical ore reserve estimation.* Amsterdam: Elsevier.

Davis, M. W. (1987). Production of conditional simulations via the LU triangular decomposition of the covariance matrix. *Mathematical Geology, 19*(2), 91–98.

Deutsch, C. V., & Journel, A. G. (1992). *GSLIB: Geostatistical Software Library and User's Guide.* New York: Oxford University Press.

Demyanov, V., Foresti, L., Christie, M., & Kanevski, M. (2011). Reservoir modelling with feature selection: A kernel learning approach. In *Proceedings of SPE Reservoir Simulation Symposium.* doi:10.2118/141510-MS.

Dimitrakopoulos, R., & Luo, X. (2004). Generalized sequential Gaussian simulation on group size ν and screen-effect approximations for large field simulations. *Mathematical Geology, 36*(5), 567–591.

Dimitrakopoulos, R., Mustapha, H., & Gloaguen, E. (2010). High-order statistics of spatial random fields: Exploring spatial cumulants for modeling complex non-Gaussian and non-linear phenomena. *Mathematical Geosciences, 42,* 65–99.

Doyen, P. M. (2007). *Seismic reservoir characterization.* Madrid: EAGE.

Dubrule, O. (2003). *Geostatistics for seismic data integration in earth models.* Tulsa, OK: SEG/EAGE Distinguished Instructor Short Course Number 6.

Fatti, J. L., Smith, G. C., Vail, P. J., Strauss, P. J., & Levitt, P. R. (1994). Detection of gas in sandstone reservoirs using AVO analysis: A 3-D seismic case history using the geostack technique. *Geophysics 59* (9) (September): 1362–1376. doi:10.1190/1.1443695

Filippova, K., Kozhenkov, A., & Alabushin, A. (2011). Seismic inversion techniques: Choice and benefits. *First Break, 29*(May), 103–114.

Francis, A. M. (2006). Understanding stochastic inversion: Part 1. *First Break, 24*(November), 79–84.

Froidevaux, R. (1993). Probability field simulation. In A. Soares (Ed.), *Geostatistics Tróia'92* (Vol. 1, pp. 73–84). Dordrecht: Kluwer.

Galli, A., Beucher, H., Le Loc'h, G., Doligez, B., & The Heresim Group. (1994). The pros and cons of the truncated Gaussian method. In M. Armstrong & P. Dowd (Eds.), *Geostatistical simulations: Proceedings of the geostatistical workshop, Fontainebleau, France* (pp. 217–233). Dordrecht: Kluwer Academic Publishers.

Gao, G., Abubakar, A., & Habashy, T. (2010). Simultaneous joint petrophysical inversion of electromagnetic and seismic measurements. *SEG Annual Meeting, 2010,* 2799–2804.

Gómez-Hernández, J. J., & Srivastava, R. H. (1990). ISIM3D: An ANSI-C three-dimensional multiple indicator conditional simulation program. *Computers & Geosciences, 16*(4), 395–440.

Gómez-Hernández, J. J., & Journel, A. G. (1993). Joint sequential simulation of multi Gaussian field, *Geostatistics Troia'92* (pp. 85–94). Dordrecht: Kluwer.

Goovaerts, P. (1997). *Geostatistics for natural resources evaluation.* New York: Oxford University Press.

Grana, D., & Della Rossa, E. (2010). Probabilistic petrophysical-properties estimation integrating statistical rock physics with seismic inversion. *Geophysics 75*(3): O21–O37. doi:10.1190/1.3386676

Grana, D., Mukerji, T., Dvorkin, J., & Mavko, G. (2012). Stochastic inversion of facies from seismic data based on sequential simulations and probability perturbation method. *Geophysics 77*(4): M53–M72. doi:10.1190/geo2011-0417.1

Grijalba-Cuenca A., & Torres-Verdin C. (2000). Geostatistical inversion of 3D seismic data to extrapolate wireline petrophysical variables laterally away from the well. In *SPE 63283.* Texas: Society of Petroleum Engineers.

Haas, A., & Dubrule, O. (1994). Geostatistical inversion—a sequential method of stochastic reservoir modeling constrained by seismic data. *First Break, 12,* 561–569.

Hoffman, T., & Caers, J. (2005). Regional probability perturbations for history matching. *Journal of Petroleum Science and Engineering, 46*(1–2), 53–71.

Horta, A., & Soares, A. (2010). Direct sequential co-simulation with joint probability distributions. *Mathematical Geosciences 42*(3): 269–92. doi:10.1007/s11004-010-9265-x

Hoversten, G. M., Cassassuce, F., Gasperikova, E., Newman, G. A., Chen, J., Rubin, Y., et al. (2006). Direct reservoir parameter estimation using joint inversion of marine seismic AVA and CSEM data. *Geophysics 71*(3): C1–C13. doi:10.1190/1.2194510

Hu, L. (2000). Gradual deformation and iterative calibration of Gaussian-related stochastic models. *Mathematical Geology, 32,* 87–108.

Hu, L., Blanc, G., & Noetinger, B. (2001). Gradual deformation and iterative calibration of sequential simulations. *Mathematical Geology, 33,* 475–489.

Iaco, S., & Maggio, S. (2011). Validation techniques for geological patterns simulations based on variogram and multiple-point statistics. *Mathematical Geosciences 43*(4) (March 31): 483–500. doi:10.1007/s11004-011-9326-9

Isaaks, E. H., & Srivastava, R. M. (1989). *An introduction to applied geostatistics*. New York: Oxford University Press.

Journel, A. G. (1974). Geostatistics for conditional simulation of orebodies. *Economic Geology, 69*, 673–680.

Journel, A. G. (1989). Fundamentals of Geostatistics in Five Lessons. In *Short Course in Geology 8*. Washington D.C.: American Geophysical Union.

Journel, A. G. (1994). Modeling uncertainty: Some conceptual thoughts. In R. Dimitrakopoulos (Ed.), *Geostatistics for next century*. Netherlands: Springer

Journel, A. G., & Huijbreghts, Ch. (1978). *Mining geostatistics*. New York: Academic Press.

Journel, A. G., & Alabert, F. (1988). Focusing on spatial connectivity of extreme-valued attributes: Stochastic indicator models of reservoir heterogeneities. In *SPE Paper 18324*. Texas: Society of Petroleum Engineers.

Journel, A. G., & Xu, W. (1994). Posterior identification of histograms conditional to local data. *Mathematical Geology, 24*(2), 149–160.

Kashib, T., & Srinivasan, S. (2006). A probabilistic approach to integrating dynamic data in reservoir models. *Journal of Petroleum Science and Engineering, 50*(3–4), 241–257.

Kleingeld, W. J., & Krige, D. G. (Eds.). (2001). *Geostatistics Cape Town'2000* (Vol. 1 and 2). South Africa: Geostatistical Association of South Africa.

Krige, D. G. (1951). A statistical approach to some basic mine valuation problems on the Witwatersrand. *Journal of the Southern African Institute of Mining and Metallurgy, 52*(6), 119–139.

Lancaster, S., & Whitcombe, D. (2000). Fast-Track Coloured Inversion. In: *SEG Expanded Abstracts* (pp. 3–6).

Lantuejoul, Ch. (2002). *Geostatistical simulations*. Springer, Berlin Heidelberg: Models and Algrithms.

Law, A. M., & Kelton, W. D. (1991). *Simulation Modeling and Analysis*. NY: McGraw-Hill.

Leuangthong, O., & Deutsch, C. V. (Eds.). (2004). *Geostatistics Banff*. Netherlands: Springer.

Lindseth, R. O. (1979). Synthetic sonic logs—a process for stratigraphic interpretation. *Geophysics, 44*, 3–26.

Liu, Y., & Journel, A. G. (2009). A package for geostatistical integration of coarse and fine scale data. *Computers & Geosciences 35*(3) (March): 527–547. doi:10.1016/j.cageo.2007.12.015

Lee, J., & Mukerji, T. (2013). The Stanford VI-E reservoir : A synthetic data set for joint seismic-em time-lapse monitoring algorithms. In *Stanford Center for Reservoir Forecasting Meeting* (pp. 1–53).

Le Loc'h, G., Beucher, H., Galli, A., & Doligez, B. (1994). Improvement in the truncated Gaussian method: Combining several Gaussian functions. In *Proceedings of ECMOR IV, Fourth European Conference on the Mathematics of Oil Recovery*.

Ma, Xin-Quan. (2002). Simultaneous inversion of prestack seismic data for rock properties using simulated annealing. *Geophysics 67*(6): 1877–1885. doi:10.1190/1.1527087

Mallet, J.L. 2002. *Geomodelling*, Oxford : Oxford University Press.

Mallet, J. L. (2004). Space-time mathematical framework for sedimentary geology. *Mathematical Geology, 36* (1), 1–31.

Mallick, S. (1995). Model based inversion of amplitude variations with offset data using a genetic algorithm. *Geophysics 60*(4) (July): 939–954. doi:10.1190/1. 1443860

Mallick, S. (1999). Some practical aspects of prestack waveform inversion using a genetic algorithm: An example from the East Texas Woodbine gas sand. *Geophysics, 64*(2), 326–336.

Marcotte, D. (1991). Cokriging with Matlab. *Computers & Geosciences, 18*(9), 1265–1280.

Mariethoz, G., Renard, P., & Straubhaar, J. (2010). The direct sampling method to perform multiple-point geostatistical simulations. *Water Resources Research, 46*(11): 1–14. doi:10.1029/2008WR007621

Mariethoz, G., & Caers, J. (2014). *Multiple-point geostatistics: Stochastic modelling with training images*. Wiley-Blackwell, Hoboken, NJ, 376 pp.

Marques, C., Azevedo, L., Demyanov, V., Soares, A., & Christie, M. (2015). Multiscale geostatistical history matching using block direct sequential simulation. In: *Petroleum Geostatistics 2015*. France: Biarritz

Mata-Lima, H. (2008). Reservoir characterization with iterative direct sequential co-simulation: Integrating fluid dynamic data into stochastic model. *Journal of Petroleum Science and Engineering 62*(3–4) (September): 59–72. doi:10.1016/j.petrol.2008.07.003

Matheron, G. (1965). *Les Variables Régionalisées et Leur Estimation*.

Matheron, G. (1974). *Random sets and integral geometry*. New York: Wiley Interscience.

Matheron, G. (1978). *Estimating and choosing: An essay on probability in practice*. Berlin, Germany: Springer.

Mavko, G., Mukerji, T., & Dvorkin, J. (2003). *The rock physics handbook*. Cambridge: Cambridge University Press.

Morris, H., Hardy, B., Efthymiou, E., & Kearney, T. (2011). Rock physics and reservoir characterization of a calcitic-dolomitic sandstone reservoir. *First Break, 29*(June), 71–79.

Mukerji, T., Avseth, P., Mavko, G., Takahashi, I., & González, E. F. (2001). Statistical rock physics: Combining rock physics, information theory, and geostatistics to reduce uncertainty in seismic reservoir characterization. *The Leading Edge 20*(3) (March): 313–319. doi:10.1190/1.1438938

Mukerji, T., Jørstad, A., Avseth, P., Mavko, G., & Granli, J. R. (2001). Mapping lithofacies and pore-fluid probabilities in a North Sea reservoir: Seismic inversions and statistical rock physics. *Geophysics, 66*(4), 988–1001.

Muge, F. (1982). *As Funções de recuperação Globais com o Instrumento de Planeamento Mineiro* (Ph.D. Thesis). IST, Portugal.

Myers, D. E. (1982). Matrix formulation of co-kriging: Jour. *Mathematical Geology, 14*(3), 249–257.

Myers, D. E. (1984). Co-kriging-new development, In G. Verly et al., (Eds.), *Geostatistics for natural resources characterization* (pp. 295–305). Dordrecht: D. Reidel Publishing.

Nunes, R., Soares, A., Neto, G. S., Dillon, L., Guerreiro, L., Caetano, H., et al. (2012). Geostatistical inversion of prestack seismic data. In *Ninth International Geostatistics Congress* (pp. 1–8). Oslo, Norway.

Oliver, D. S., & Chen, Y. (2011). Recent progress on reservoir history matching: a review (February 2010), 185–221. doi:10.1007/s10596-010-9194-2.

Oliver, D. S., Reynolds, A., & Liu, N. (2008). *Inverse theory for petroleum reservoir characterization and history matching*. Cambridge University Press.

Omre, H., Solna, K., & Tjelmeland, H. (1993). Simulation of random functions on large lattices, In A. Soares, (Ed.), *Geostatistics Troia '92* (pp. 179–199). Dordrecht: Kluwer Acrs.

Ortiz, J. M., & Emery, X. (Eds.). (2008). *Geostatistics Santiago 2008*. Netherlands: Springer.

Pyrcz, M. J., & Deutsch, C. (2002). *Geostatistical reservoir modeling* (2nd ed.). New York: Oxford University Press.

Renard, P., & Allard, D. (2013). Connectivity metrics for subsurface flow and transport. *Advances in Water Resources, 51*, 168–196.

Richmond, A., & Dimitrakopoulos, R. (2005). *Multi-scale stochastic modeling of ore textures at the George Fisher mine* (p. 8). Queensland, Australia: CIM Bulletin, May.

Ripley, B. D. (1987). *Stochastic Simulation*. New York: Wiley.

Robinson, G. (2001). Stochastic seismic inversion applied to reservoir characterization. *CSEG Recorder, 26*(1), 38–40.

Russell, B. H. (1988). *Introduction to seismic inversion methods*. Tulsa: SEG.

Russell, B., & Hampson, D. (1991). Comparison of poststack seismic inversion methods. In *61st Annual International Meeting, SEG, Expanded Abstracts* (pp. 876–878).

Rutherford, S. R., & Williams, R. H. (1989). Amplitude-versus-offset variations in gas sands. *Geophysics, 54*(2), 680–688.

Samper, F. J., & Carrera, J. (1990). *Geoestadística*. Centro Internacional de Métodos Númericos en Ingenieria, Barcelona: Aplicaciones a la Hidrogeologia Subterrânea.

Scales, J. A., & Tenorio, L. (2001). Prior information and uncertainty in inverse problems. *Geophysics, 66*(2), 389–397.

Scheidt, C., & Caers, J. (2008). Representing spatial uncertainty using distances and kernels. *Mathematical Geosciences 41*(4) (September 24): 397–419. doi:10.1007/s11004-008-9186-0

Sen, M. K., & Stoffa, P. L. (1991). Nonlinear one dimensional seismic waveform inversion using simulated annealing. *Geophysics 56*(10) (October): 1624–1638. doi:10.1190/1.1442973

Serra, J. (1982). *Image analysis and mathematical morphology*. London, New York: Academic Press.

Shuey, R. T. 1985. A simplification of the Zoeppritz Equations. *Geophysics 50*(4) (April): 609–614. doi:10.1190/1.1441936

Simm, R., & Bacon, M. (2014). *Seismic amplitude An interpreter's handbook*. UK: Cambridge University Press.

Smith, T. M., Sondergeld, C. H., & Rai, C. S. (2003). Gassmann fluid substitutions: A tutorial. *Geophysics, 68*(2), 430–440.

Soares, A. (1988). Conditional simulation of indicator data. Case study of a multiseam coal deposits. In Chung et al. (Eds.), *Quantitative analysis of mineral and energy resources* (pp. 375–384), Dordrecht: Reidel Pub.

Soares, A. (1992). Geostatistical estimation of multi/phase structures. *Mathematical Geology, 24*(2), 149–160.

Soares, A. (Ed.). 1993. *Geostatistics TROIA'92* (Vol. 1 and 2). Dordrecht: Kluwer Academic Publishers.

Soares, A. (2001). Direct Sequential Simulation and Cosimulation. *Mathematical Geology, 33*(8), 911–926.

Soares, A. (2006). *Geoestatística Para as Ciências Da Terra e Do Ambiente*. Lisboa, Portugal: IST Press.

Soares, A., Diet, J.D., & Guerreiro, L. (2007). Stochastic inversion with a global perturbation method. Petroleum Geostatistics, Cascais, Portugal: EAGE, September 2007: 10–14.

Sousa, A. J. (1989). Geostatistical data analysis—an application to ore typology. In M. Armstrong, (Ed.), *Geostatistics: Vol. 2* (pp. 295–308). Dordrecht: Kluwer.

Srivastava, R. M. (1992). *Reservoir characterization with probability field simulation*, SPE paper 24753. Texas: Society of Petroleum Engineers.

Srivastava, R. M. (1995). An overview of stochastic methods for reservoir characterization. In J. M. Yarus and R. L. Chambers (Eds.), *Stochastic modeling and geostatistics: Principles, methods, and case studies: AAPG computer applications in geology: Vol. 3* (pp. 1–16).

Strebelle, S. (2002). Conditional simulation of complex geological structures using multiple-point statistics. *Mathematical Geology, 34*(1), 1–21.

Suro-Perez, V., & Journel, A. (1990). Simulation of lithofacies. In Guerrilot et al. (Eds.), *Proceedings of the 2nd European Conference on Mathematics of Oil Recovery* (pp. 3–10). Paris: Tech. Pub., Technip.

Switzer, P. (1977). Estimation of spatial distribution from point sources with applications to air pollution measurements. *Bulletin of the Institute of Statistics, XLVII*(1), 123–137.

Sun, Q., Eissa, M.A., Castagna, J.P., Cersosimo, D., Sun, S., & Deker, C. (2001). Porosity from artificial neural network inversion from Bermejo Field, Ecuador. In *71st Annual International Meeting: Society of Exploration Geophysicists* (pp. 734–737)

Suzuki, S., & Caers, J. (2008). A distance-based prior model parameterization for constraining solutions of spatial inverse problems. *Mathematical Geosciences* *40*(4) (March 12): 445–469. doi:10.1007/s11004-008-9154-8

Tarantola, A. (2005). *Inverse problem theory*. SIAM.

Tompkins, M. J., Fernández Martínez, J. L., Alumbaugh, D. L., & Mukerji, T. (2011). Scalable uncertainty estimation for nonlinear inverse problems using parameter reduction, constraint mapping, and geometric sampling: Marine controlled-source. *Geophysics,* *76*(4), 263–281.

Verly, G. 1993. Sequential Gaussian co-simulation: A simulation method integrating several types of information. In A. Soares, (Ed.), *Geostatistics Troia '92* (Vol. 1, pp. 543–554). Dordrecht, The Netherlands: Kluwer Academic Publishers.

Ventsel, H. (1973). *Théorie des Probabilités*. Edition Mir, Moscou.

Verly, G., David, M., Journel, A. G., & Marechal, A. (Eds.). (1984). *Geostatistics for natural resources characterization*. Dordrecht, Holland: D. Reidel Publishing.

Wackernagel, H. (1995). *Multivariate geostatistics, an introduction with applications*. Berlin: Springer.

Waxman, M. H., & Smits, M. J. L. (1968). Electrical conduction in oil-bearing sands. *Petroleum Engineers Journal, 8*(2), 107–122.

Xu, S., & Payne, M. A. (2009). Modeling elastic properties in carbonate rocks. *The Leading Edge* (January): 66–74.

Xu, W., Tran, T., Srivastava, M., & Journel, A. G. (1992). Integrating seismic data in reservoir modelling: The collocated co-kriging alternative, SPE paper 247242. Texas: Society of Petroleum Engineers.

Index

© Springer International Publishing AG 2017
L. Azevedo and A. Soares, *Geostatistical Methods for Reservoir Geophysics*,
Advances in Oil and Gas Exploration & Production, DOI 10.1007/978-3-319-53201-1

Printed in the United States
By Bookmasters